T0291870

CAMBRIDGE LIBRARY COLLECTION

Books of enduring scholarly value

Cambridge

The city of Cambridge received its royal charter in 1201, having already been home to Britons, Romans and Anglo-Saxons for many centuries. Cambridge University was founded soon afterwards and celebrates its octocentenary in 2009. This series explores the history and influence of Cambridge as a centre of science, learning, and discovery, its contributions to national and global politics and culture, and its inevitable controversies and scandals.

A History of the Study of Mathematics at Cambridge

For centuries, Cambridge University has attracted some of the world's greatest mathematicians. This 1889 book gives a compelling account of how mathematics developed at Cambridge from the middle ages to the late nineteenth century, from the viewpoint of a leading scholar based at Trinity College who was closely involved in teaching the subject. The achievements of notable individuals including Newton and his school are set in the context of the history of the university, its sometimes uneasy relationship with the town community, the college system, and the origin and growth of the mathematical tripos.

Cambridge University Press has long been a pioneer in the reissuing of out-of-print titles from its own backlist, producing digital reprints of books that are still sought after by scholars and students but could not be reprinted economically using traditional technology. The Cambridge Library Collection extends this activity to a wider range of books which are still of importance to researchers and professionals, either for the source material they contain, or as landmarks in the history of their academic discipline.

Drawing from the world-renowned collections in the Cambridge University Library, and guided by the advice of experts in each subject area, Cambridge University Press is using state-of-the-art scanning machines in its own Printing House to capture the content of each book selected for inclusion. The files are processed to give a consistently clear, crisp image, and the books finished to the high quality standard for which the Press is recognised around the world. The latest print-on-demand technology ensures that the books will remain available indefinitely, and that orders for single or multiple copies can quickly be supplied.

The Cambridge Library Collection will bring back to life books of enduring scholarly value across a wide range of disciplines in the humanities and social sciences and in science and technology.

A History of the Study of Mathematics at Cambridge

WALTER WILLIAM ROUSE BALL

CAMBRIDGE UNIVERSITY PRESS

Cambridge New York Melbourne Madrid Cape Town Singapore São Paolo Delhi

Published in the United States of America by Cambridge University Press, New York

www.cambridge.org
Information on this title: www.cambridge.org/9781108002073

This edition first published 1889
This digitally printed version 2009

ISBN 978-1-108-00207-3

A HISTORY

OF THE STUDY OF

MATHEMATICS AT CAMBRIDGE.

London: C. J. CLAY AND SONS,
CAMBRIDGE UNIVERSITY PRESS WAREHOUSE,
AVE MARIA LANE.

Cambridge: DEIGHTON, BELL, AND CO.
Leipzig: F. A. BROCKHAUS.

A HISTORY

OF THE STUDY OF

MATHEMATICS AT CAMBRIDGE

BY

W. W. ROUSE BALL,

FELLOW AND LECTURER OF TRINITY COLLEGE, CAMBRIDGE;
AUTHOR OF A HISTORY OF MATHEMATICS.

Cambridge:
AT THE UNIVERSITY PRESS.
1889

Cambridge:

PRINTED BY C. J. CLAY, M.A. AND SONS,
AT THE UNIVERSITY PRESS.

PREFACE.

THE following pages contain an account of the development of the study of mathematics in the university of Cambridge, and the means by which proficiency in that study was at various times tested. The general arrangement is as follows.

The first seven chapters are devoted to an enumeration of the more eminent Cambridge mathematicians, arranged chronologically. I have in general contented myself with mentioning the subject-matter of their more important works, and indicating the methods of exposition which they adopted, but I have not attempted to give a detailed analysis of their writings. These chapters necessarily partake somewhat of the nature of an index. A few remarks on the general characteristics of each period are given in the introductory paragraphs of the chapter devoted to it; and possibly for many readers this will supply all the information that is wanted.

The following chapters deal with the manner in which at different times mathematics was taught, and the means by which proficiency in the study was tested. The table of contents will shew how they are arranged. Some knowledge of the constitution, organization and general history of the university is, in my opinion, essential to any who would understand the way in which mathematics was introduced into the university curriculum, and its relation to other departments of study. I have therefore added in chapter XI. (as a sort of appendix) a very brief sketch of the general history of the university for any of my readers who may not be acquainted with the larger works which deal with that subject. I hope that the addition of that chapter and of the similar chapter dealing with the organization of studies in the mediæval university will sufficiently justify me in the use in the earlier chapters of various technical words, such as regents, caput, tripos, prævaricator, &c.

I have tried to give references in the footnotes to the authorities on which I have mainly relied. In the few cases where no reference is inserted, I have had to compile my account from various sources. Of the numerous dictionaries of biography which I have consulted the only ones which have proved of much use are the *Biographica Britannica*, six volumes, London, 1747—66 (second edition, enlarged, letters *A* to *Fas* only, five volumes, 1778—93); the *Penny Cyclopaedia*, twenty-seven volumes, London, 1833—43; J. C. Poggendorff's *Biographisch-Literarisches Handwörterbuch zur Geschichte der exacten Wissenschaften*, two volumes, Leipzig, 1863; and the new *Dictionary of national biography*, which at present only contains references to those whose names commence with one of the early letters of the alphabet. To these four works I have been constantly indebted: I have found them almost always reliable, and very useful, not only where no other accounts were available, but also for the verification of such biographical notes as I had given, and often for the addition of other details to them.

No one who has not been engaged in such a work can imagine how difficult it is to settle many a small point, or how persistently mistakes if once printed are reproduced in every subsequent account. In spite of the care I have taken I have no doubt that there are some omissions and errors in the following pages; and I shall thankfully accept notices of additions or corrections which may occur to any of my readers.

W. W. ROUSE BALL.

TRINITY COLLEGE, CAMBRIDGE.
May, 1889.

TABLE OF CONTENTS.

Chapter I. Mediæval mathematics.

Chapter II. The mathematics of the renaissance.

Chapter III. The commencement of modern mathematics.

Chapter IV. The life and works of Newton.

Chapter V. The rise of the Newtonian school.

Chapter VI. The later Newtonian school.

Chapter VII. The analytical school.

Chapter VIII. The organization and subjects of education.

Chapter IX. The exercises in the schools.

Chapter X. The mathematical tripos.

Chapter XI. Outlines of the history of the university.

ERRATA.

Page 14, line 3. After *under* insert *the.*

Page 34, line 8. For *powers* read *power.*

Page 38, lines 3 and 5. For *Bulialdus* read *Bullialdus.*

Page 91, line 12. For *seventeenth* read *eighteenth.*

Page 92, line 4 from end, and page 95, line 5 from end. For *Lahire* read *La Hire.*

Page 115, line 12. For *His* read *Cavendish's.*

Page 183, line 20. For *T. Bowstead* read *Joseph Bowstead.*

CHAPTER I.

MEDIAEVAL MATHEMATICS.[1]

THE subject of this chapter is a sketch of the nature and extent of the mathematics read at Cambridge in the middle ages. The external conditions under which it was carried on are briefly described in the first section of chapter VIII. It is only after considerable hesitation that I have not incorporated that section in this chapter; but I have so far isolated it as to render it possible, for any who may be ignorant of the system of education in a mediæval university, to read it if they feel so inclined, before commencing the history of the development of mathematics at Cambridge.

The period with which I am concerned in this chapter begins towards the end of the twelfth century, and ends with the year 1535. For the history during most of this time of mathematics at Cambridge we are obliged to rely largely on inferences from the condition of other universities. I shall therefore discuss it very briefly referring the reader to the works mentioned below[1] for further details.

[1] Besides the authorities alluded to in the various foot-notes I am indebted for some of the materials for this chapter to *Die Mathematik auf den Universitäten des Mittelalters* by H. Suter, Zurich, 1887; *Die Geschichte des mathematischen Unterrichtes im deutschen Mittelalter bis* 1525, by M. S. Günther, Berlin, 1887; and *Beiträge zur Kenntniss der Mathematik des Mittelalters*, by H. Weissenborn, Berlin, 1888.

Throughout the greater part of this period a student usually proceeded in the faculty of arts; and in that faculty he spent the first four years on the study of the subjects of the trivium, and the next three years on those of the quadrivium. The trivium comprised Latin grammar, logic, and rhetoric; and I have described in chapter VIII. both how they were taught and the manner in which proficiency in them was tested. It must be remembered that students while studying the trivium were treated exactly like school-boys, and regarded in the same light as are the boys of a leading public school at the present time. The title of bachelor was given at the end of this course. A bachelor occupied a position analogous to that of an undergraduate now-a-days. He was required to spend three years in the study of the quadrivium, the subjects of which were mathematics and science. These were divided in the Pythagorean manner into numbers absolute or arithmetic, numbers applied or music, magnitudes at rest or geometry, and magnitudes in motion or astronomy. There was however no test that a student knew anything of the four subjects last named other than his declaration to that effect, and in practice most bachelors left them unread. The degree of master was given at the end of this course.

The quadrivium during **the twelfth and the first half of the thirteenth century**, if studied at all, probably meant about as much science as was to be found in the pages of Boethius, Cassiodorus, and Isidorus. The extent of this is briefly described in the following paragraphs.

The term arithmetic was used in the Greek sense, and meant the study of the properties of numbers; and particularly of ratio, proportion, fractions, and polygonal numbers. It did not include the art of practical calculation, which was generally performed on an abacus; and though symbols were employed to express the result of any numerical computation they were not used in determining it.

The geometry was studied in the text-books either of

Boethius or of Gerbert[1]. The former work, which was the one more commonly used, contains the enunciations of the first book of Euclid, and of a few selected propositions from the third and fourth books. To shew that these are reliable, demonstrations of the first three propositions of the first book are given in an appendix. Some practical applications to the determination of areas were usually added in the form of notes. Even this was too advanced for most students. Thus Roger Bacon, writing towards the close of the thirteenth century, says that at Oxford, there were few, if any, residents who had read more than the definitions and the enunciations of the first five propositions of the first book. In the pages of Cassiodorus and Isidorus a slight sketch of geography is included in geometry.

The two remaining subjects of the quadrivium were music and astronomy. The study of the former had reference to the services of the Church, and included some instruction in metre. The latter was founded on Ptolemy's work, and special attention was supposed to be paid to the rules for finding the moveable festivals of the Church; but it is probable that in practice it generally meant the art of astrology.

By **the middle of the thirteenth century** anyone who was really interested in mathematics had a vastly wider range of reading open to him, but whether students at the English universities availed themselves of it is doubtful.

The mathematical science of modern Europe dates from the thirteenth century, and received its first stimulus[2] from the Moorish schools in Spain and Africa, where the Arab works on arithmetic and algebra, and the Arab translations of Euclid, Archimedes, Apollonius, and Ptolemy were not uncommon. It will be convenient to give here an outline of

[1] Prof. Weissenborn thinks that neither of these books was written by its reputed author, and assigns them both to the eleventh century. This view is not however generally adopted.

[2] For further details of the introduction of Arab science into Europe, see chapter x. of my *History of mathematics*, London, 1888.

the introduction of the Arab geometry and arithmetic into Europe.

First, for the geometry. As early as 1120 an English monk, named Adelhard (of Bath), had obtained a copy of a Moorish edition of the *Elements* of Euclid; and another specimen was secured by Gerard of Cremona in 1186. The first of these was translated by Adelhard, and a copy of this fell into the hands of Giovanni Campano or Campanus, who in 1260 reproduced it as his own. The first printed edition was taken from it and was issued by Ratdolt at Venice in 1482: of course it is in Latin. This pirated translation was the only one generally known until in 1533 the original Greek text was recovered[1]. Campanus also issued a work founded on Ptolemy's astronomy and entitled the *Theory of the planets*.

The earliest explanation of the Arabic system of arithmetic and algebra, which had any wide circulation in Europe, was that contained in the *Liber abbaci* issued in 1202 by Leonardo of Pisa. In this work Leonardo explained the Arabic system of numeration by means of nine digits and a zero; proved some elementary algebraical formulæ by geometry, as in the second book of Euclid; and solved a few algebraical equations. The reasoning was expressed at full length in words and without the use of any symbols. This was followed in 1220 by a work in which he shewed how algebraical formulæ could be applied to practical geometrical problems, such as the determination of the area of a triangle in terms of the lengths of the sides.

Some ten or twelve years later, circ. 1230, the emperor Frederick II. engaged a staff of Jews to translate into Latin all the Arab works on science which were obtainable; and manuscript transcripts of these were widely distributed. Most of the mediæval editions of the writings of Ptolemy, Archimedes, and Apollonius were derived from these copies.

One branch of this science of the Moors was almost at once adopted throughout Europe. This was their arithmetic, which

[1] See p. 23, hereafter; and also the article *Eucleides*, by A. De Morgan, in Smith's *Dictionary of Greek and Roman biography*, London, 1849.

was commonly known as algorism, or the art of Alkarismi, to distinguish it from the arithmetic founded on the work of Boethius. From the middle of the thirteenth century this was used in nearly all mathematical tables, whether astronomical, astrological, or otherwise. It was generally employed for trade purposes by the Italian merchants at or about the same time, and from them spread through the rest of Europe. It would however seem that this rapid adoption of the Arabic numerals and arithmetic was at least as largely due to the calculators of calendars as to merchants and men of science. Perhaps the oriental origin of the symbols gave them an attractive flavour of magic, but there seem to have been very few almanacks after the year 1300 in which an explanation of the system was not included.

The earliest lectures on the subjects of algebra and algorism at any university, of which I can find mention, are some given by Holywood, who is perhaps better known by the latinized name of Sacrobosco. *John de Holywood* was born in Yorkshire and educated at Oxford, but after taking his master's degree he moved to Paris and taught there till his death in 1244 or 1246. His work on arithmetic[1] was for many years a standard authority. He further wrote a treatise on the sphere, which was made public in 1256: this had a wide circulation, and indicates how rapidly a knowledge of mathematics was spreading. Besides these, two pamphlets by him, entitled respectively *De computo ecclesiastico* and *De astrolabio*, are still extant.

Towards **the end of the thirteenth century** a strong effort was made to introduce this science, as studied in Italy, into the curriculum of the English universities. This was due to *Roger Bacon*[2]. Bacon, who was educated at Oxford and Paris

[1] This was printed at Paris in 1496 under the title *De algorithmo;* and has been reissued in Halliwell's *Rara mathematica*, London, second edition, 1841. See also pp. 13—15 of *Arithmetical books*, by A. De Morgan, London, 1847.

[2] See *Roger Bacon, sa vie, ses ouvrages...* by E. Charles, Paris, 1861; and *Roger Bacon, eine Monographie,* by Schneider, Augsburg, 1873. The first of these is very eulogistic, the latter somewhat severely critical. An

and taught at both universities, declared that divine mathematics was not only the alphabet of all philosophy, but should form the foundation of all liberal education, since it alone could fit the student for the acquirement of other knowledge, and enable him to detect the false from the true. He urged that it should be followed by linguistic or scientific studies. These seem also to have been the views of Grosseteste, the statesmanlike bishop of Lincoln. But the power of the schoolmen in the universities was too strong to allow of such a change, and not only did they prevent any alteration of the curriculum but even the works of Bacon on physical science (which might have been included in the quadrivium) were condemned as tending to lead men's thoughts away from the problems of philosophy. It is clear, however, that henceforth a student, who was desirous of reading beyond the narrow limits of the schools, had it in his power to do so: and if I say nothing more about the science of this time it is because I think it probable that no such students were to be found in Cambridge.

The only notable English mathematician in **the early half of the fourteenth century** of whom I find any mention is *Thomas Bradwardine*[1], archbishop of Canterbury. Bradwardine was born at Chichester about 1290. He was educated at Merton College, Oxford, and subsequently lectured in that university. From 1335 to the time of his death he was chiefly occupied with the politics of the church and state: he took a prominent part in the invasion of France, the capture of Calais, and the victory of Cressy. He died at Lambeth in 1349. His mathematical works, which were probably written when he was at Oxford, are (i) the *Tractatus de proportionibus*, printed at Paris in 1495; (ii) the *Arithmetica speculativa*, printed

account of his life by J. S. Brewer is prefixed to the Rolls Series edition of the *Opera inedita*, London, 1859.

[1] See vol. IV. of the *Lives of the Archbishops of Canterbury*, by W. F. Hook, London, 1860—68; see also pp. 480, 487, 521—24 of the *Aperçu historique sur...géométrie* by M. Chasles (first edition).

at Paris in 1502; (iii) the *Geometria speculativa*, printed at Paris in 1511; and (iv) the *De quadratura circuli*, printed at Paris in 1495. They probably give a fair idea of the nature of the mathematics then read at an English university.

By the middle of this century Euclidean geometry (as expounded by Campanus) and algorism were fairly familiar to all professed mathematicians, and the Ptolemaic astronomy was also generally known. About this time the almanacks began to add to the explanation of the Arabic symbols the rules of addition, subtraction, multiplication, and division, "de algorismo." The more important calendars and other treatises also inserted a statement of the rules of proportion, illustrated by various practical questions; such books usually concluded with algebraic formulæ (expressed in words) for most of the common problems of commerce. Of course the fundamental rules of this algorism were not strictly proved—that is the work of advanced thought—but it is important to note that there was some discussion of the principles involved.

I should add that next to the Italians the English took the most prominent part in the early development and improvement of algorism[1], a fact which the backward condition of the country makes rather surprising. Most merchants continued however to keep their accounts in Roman numerals till about 1550, and monasteries and colleges till about 1650: though in both cases it is probable that the processes of arithmetic were performed in the algoristic manner. No instance in a parish register of a date or number being written in Arabic numerals is known to exist before the seventeenth century.

In **the latter half of the fourteenth century** an attempt was made to include in the quadrivium these new works on the elements of mathematics. The stimulus came from Paris, where a statute to that effect was passed in 1366, and a year or two later similar regulations were made at Oxford and Cam-

[1] An English treatise by John Norfolk, written about 1340, is still extant. It was printed in 1445 and reissued by Halliwell in his *Rara mathematica*, London, second edition, 1841.

bridge; unfortunately no text-books[1] are mentioned. We can however form a reasonable estimate of the range of mathematical reading required, by looking at the statutes of the universities of Prague founded in 1350, of Vienna founded in 1364, and of Leipzig founded in 1389[2].

By the statutes of Prague[3], dated 1384, candidates for the bachelor's degree were required to have read Holywood's treatise on the sphere, and candidates for the master's degree to be acquainted with the first six books of Euclid, optics, hydrostatics, the theory of the lever, and astronomy. Lectures were actually delivered on arithmetic, the art of reckoning with the fingers, and the algorism of integers; on almanacks, which probably meant elementary astrology; and on the *Almagest*, that is on Ptolemaic astronomy. There is however some reason for thinking that mathematics received there far more attention than was then usual at other universities.

At Vienna in 1389 the candidate for a master's degree was required[4] to have read five books of Euclid, common perspective, proportional parts, the measurement of superficies, and the *Theory of the planets*. The book last named is the treatise by Campanus, which was founded on that by Ptolemy. This was a fairly respectable mathematical standard, but I would remind the reader that there was no such thing as "plucking" in a mediæval university. The student had to keep an act or give a lecture on certain subjects, but whether he did it well or badly he got his degree, and it is probable that it was only the few students whose interests were mathematical who really mastered the subjects mentioned above.

[1] See p. 81 of *De l'organisation de l'enseignement...au moyen âge* by C. Thurot, Paris, 1850.

[2] The following account is taken from *Die Geschichte der Mathematik*, by H. Hankel, Leipzig, 1874.

[3] See vol. I. pp. 49, 56, 77, 83, 92, 108, 126, of the *Historia universitatis Pragensis*, Prag, 1830.

[4] See vol. I. p. 237 of the *Statuta universitatis Wiennensis* by V. Kollar, Vienna, 1839: quoted in vol. I. pp. 283 and 351 of the *University of Cambridge*, by J. Bass Mullinger, Cambridge, 1873.

At any rate no test of proficiency was imposed; and a few facts gleaned from the history of the next century tend to shew that the regulations about the study of the quadrivium were not seriously enforced. The lecture lists for the years 1437 and 1438 of the university of Leipzig (the statutes of which are almost identical with those of Prague as quoted above) are extant, and shew that the only lectures given there on mathematics in those years were confined to astrology. The records[1] of Bologna, Padua, and Pisa seem to imply that there also astrology was the only scientific subject taught in the fifteenth century, and even as late as 1598 the professor of mathematics at Pisa was required to lecture on the *Quadripartitum*, a spurious astrological work attributed to Ptolemy. According to the registers[2] of the university of Oxford the only mathematical subjects read there between the years 1449 and 1463 were Ptolemy's astronomy (or some commentary on it) and the first two books of Euclid. Whether most students got as far as this is doubtful. It would seem, from an edition of Euclid published at Paris in 1536, that after 1452 candidates for the master's degree at that university had to take an oath that they had attended lectures on the first six books of Euclid.

The only Cambridge mathematicians of **the fifteenth century** of whom I can find any mention were Holbroke, Marshall, and Hodgkins. No details of their lives and works are known. **John Holbroke**, master of Peterhouse and chancellor of the university for the years 1428 and 1429, who died in 1437, is reputed to have been a distinguished astronomer and astrologer. **Roger Marshall**, who was a fellow of Pembroke, taught mathematics and medicine; he subsequently moved to London and became physician to Edward IV. **John Hodgkins**, a fellow of King's, who died in 1485 is mentioned as a celebrated mathematician.

[1] See pp. 15, 20 of *Die Geschichte der mathematischen Facultät in Bologna* by S. Gherardi, edited by M. Kurtze, Berlin, 1871.

[2] Quoted in the *Life of bishop Smyth* (the founder of Brazenose College) by Ralph Churton, Oxford, 1800.

At **the beginning of the sixteenth century** the names of Master, Paynell, and Tonstall occur. Of these **Richard Master,** a fellow of King's, is said to have been famous for his knowledge of natural philosophy. He entered at King's in 1502, and was proctor in 1511. He took up the cause of the holy maid of Kent and was executed for treason in April, 1534. **Nicholas Paynell,** a fellow of Pembroke Hall, graduated B.A. in 1515. In 1530 he was appointed mathematical lecturer to the university. The date of his death is unknown.

Cuthbert Tonstall[1] was born at Hackforth, Yorkshire, in 1474 and died in 1559. He had entered at Balliol College, Oxford, but finding the philosophers dominant in the university (see p. 243), he migrated to King's Hall, Cambridge. We must not attach too much importance to this step for such migrations were then very common, and his action only meant that he could continue his studies better at Cambridge than at Oxford. He subsequently went to Padua, where he studied the writings of Regiomontanus and Pacioli. His arithmetic termed *De arte supputandi* was published in 1522 as a "farewell to the sciences" on his appointment to the bishopric of London. A presentation copy on vellum with the author's autograph is in the university library at Cambridge. The work is described by De Morgan in his *Arithmetical Books* as one of the best which has been written both in matter, style, and for the historical knowledge displayed. Few critics would agree with this estimate, but it was undoubtedly the best arithmetic then issued, and forms a not unworthy conclusion to the mediæval history of Cambridge. It is particularly valuable as containing illustrations of the mediæval processes of computation. Several extracts from it are given in the *Philosophy of arithmetic* by J. Leslie, second edition, Edinburgh, 1820.

That brings me to the close of the middle ages, and the above account—meagre though it is—contains all that I have

[1] See vol. I. p. 198 of the *Athenae Cantabrigienses* by C. H. and T. Cooper, Cambridge, 1858—61.

been able to learn about the extent of mathematics then taught at an English university. About Cambridge in particular I can give no details. The fact however that Tonstall and Recorde, the only two English mathematicians of any note of the first half of the sixteenth century, both migrated from Oxford to Cambridge in order to study science makes it probable that it was becoming an important school of mathematics.

CHAPTER II.

THE MATHEMATICS OF THE RENAISSANCE.

CIRC. 1535—1630.

THE close of the mediæval period is contemporaneous with the beginning of the modern world. The reformation and the revival of the study of literature flooded Europe with new ideas, and to these causes we must in mathematics add the fact that the crowds of Greek refugees who escaped to Italy after the fall of Constantinople brought with them the original works and the traditions of Greek science. At the same time the invention of printing (in the fifteenth century) gave facilities for disseminating knowledge which made these causes incomparably more potent than they would have been a few centuries earlier.

It was some years before the English universities felt the full force of the new movement, but in 1535 the reign of the schoolmen at Cambridge was brought to an abrupt end by "the royal injunctions" of that year (see p. 244). Those injunctions were followed by the suppression of the monasteries and the schools thereto attached, and thus the whole system of mediæval education was destroyed. Then ensued a time of great confusion. The number of students fell, so that the entries for the decade ending 1547 are probably the lowest in the whole seven centuries of the history of the university.

The writings of Tonstall and Recorde, and the fact that most of the English mathematicians of the time came from Cambridge seem to shew that mathematics was then regularly taught, and of course according to the statutes it still constituted the course for the M.A. degree. But it is also clear that it was only beginning to grow into an important study, and was not usually read except by bachelors, and probably by only a few of them. The chief English mathematician of this time was Recorde whose works are described immediately hereafter; but John Dee, Thomas Digges, Thomas Blundeville, and William Buckley were not undistinguished.

The period of confusion in the studies of the university caused by the break-up of the mediæval system of education was brought to an end by the Edwardian statutes of 1549 (see p. 153). These statutes represented the views of a large number of residents, and it is noticeable that they enjoined the study of mathematics as the foundation of a liberal education. Certain text-books were recommended, and we thus learn that arithmetic was usually taught from Tonstall and Cardan, geometry from Euclid, and astronomy from Ptolemy. Cosmography was still included in the quadrivium, and the works of Mela, Strabo, and Pliny are referred to as authorities on it.

The Edwardian code was only in force for about twenty years. Fresh statutes were given by Elizabeth in 1570, and except for a few minor alterations these remained in force till 1858. The commissioners who framed them excluded mathematics from the course for undergraduates—apparently because they thought that its study appertained to practical life, and had its place in a course of technical education rather than in the curriculum of a university. These opinions were generally held at that time[1] and it will be found that most of the English books on the subject issued for the following sixty or seventy years—the period comprised in this chapter—were chiefly devoted to practical applications, such as surveying, navigation, and astrology. Accordingly we find that for the next half century mathematics was more studied in London than at the universities, and it was not until it became a

[1] See for example vol. I. pp. 382—91 of the *Orationes* of Melanchthon, and the autobiography of Lord Herbert of Cherbury (born in 1581 and died in 1648) which was published in London in 1792.

science (under the influence of Wallis, Barrow, and Newton) that much attention was paid to it at Cambridge.

It must however be remembered that though under Elizabethan statutes mathematics was practically relegated to a secondary position in the university curriculum, yet it remained the statutable subject to be read for the M.A. degree. That was in accordance with the views propounded by Ramus[1] who considered that a liberal education should comprise the exoteric subjects of grammar, rhetoric, and dialectics; and the esoteric subjects of mathematics, physics, and metaphysics for the more advanced students. The exercises for the degree of master were however constantly neglected, and after 1608 when residence was declared to be unnecessary (see p. 157) they were reduced to a mere form.

I think it will be found that in spite of this official discouragement the majority of the English mathematicians of the early half of the seventeenth century were educated at Cambridge, even though they subsequently published their works and taught elsewhere.

Among the more eminent Cambridge mathematicians of the

[1] See p. 346 of *Ramus; sa vie, ses écrits, et ses opinions* by Ch. Waddington, Paris, 1855. Another sketch of his opinions is given in a monograph of which he is the subject by C. Desmaze, Paris, 1864. *Peter Ramus* was born at Cuth in Picardy in 1515, and was killed at Paris at the massacre of St Bartholomew on Aug. 24, 1572. He was educated at the university of Paris, and on taking his degree he astonished and charmed the university with the brilliant declamation he delivered on the thesis that everything Aristotle had taught was false. He lectured first at le Mans, and afterwards at Paris; at the latter he founded the first chair of mathematics. Besides some works on philosophy he wrote treatises on arithmetic, algebra, geometry (founded on Euclid), astronomy (founded on the works of Copernicus), and physics which were long regarded on the continent as the standard text-books on these subjects. They are collected in an edition of his works published at Bâle in 1569. Cambridge became the chief centre for the Ramistic doctrines, and was apparently frequented by foreign students who desired to learn his logic and system of philosophy: see vol. II. pp. 411—12 of the *University of Cambridge*, by J. Bass Mullinger, Cambridge, 1884.

latter half of the sixteenth century I should include Sir Henry Billingsley, Thomas Hill, Thomas Bedwell, Thomas Hood, Richard Harvey, John Harvey, and Simon Forman. These were only second-rate mathematicians. They were followed by Edward Wright, Henry Briggs, and William Oughtred, all of whom were mathematicians of mark: most of the works of the three last named were published in the seventeenth century.

After this brief outline of my arrangement of the chapter I return to the Cambridge mathematicians of the first half of the sixteenth century.

The earliest of these—if we except Tonstall—and the first English writer on pure mathematics of any eminence was Recorde. **Robert Recorde**[1] was born at Tenby about 1510. He was educated at Oxford, and in 1531 obtained a fellowship at All Souls' College; but like Tonstall he found that there was then no room at that university for those who wished to study science beyond the traditional and narrow limits of the quadrivium. He accordingly migrated to Cambridge, where he read mathematics and medicine. He then returned to Oxford, but his reception was so unsatisfactory that he moved to London, where he became physician to Edward VI. and afterwards to Queen Mary. His prosperity however must have been short-lived, for at the time of his death in 1558 he was confined in the King's Bench prison for debt.

His earliest work was an arithmetic published in 1540 under the title the *Grounde of artes.* This is the earliest English scientific work of any value. It is also the first English book which contains the current symbols for addition,

[1] See the *Athenae Cantabrigienses* by C. H. and T. Cooper, two vols. Cambridge, 1858 and 1863. To save repetition I may say here, once for all, that the accounts of the lives and writings of such of the mathematicians as are mentioned in the earlier part of this chapter and who died before 1609 are founded on the biographies contained in the *Athenae Cantabrigienses.*

subtraction, and equality. There are faint traces of his having used the two former as symbols of operation and not as mere abbreviations. The sign = for equality was his invention. He says he selected that particular symbol because than two parallel straight lines no two things can be more equal, but M. Charles Henry has pointed out in the *Revue archéologique* for 1879 that it is a not uncommon abbreviation for the word *est* in mediæval manuscripts, and this would seem to point to a more probable origin. Be this as it may, the work is the best treatise on arithmetic produced in that century.

Most of the problems in arithmetic are solved by the rule of false assumption. This consists in assuming any number for the unknown quantity, and if on trial it does not satisfy the given conditions, correcting it by simple proportion as in rule of three. It is only applicable to a very limited class of problems. As an illustration of its use I may take the following question. A man lived a fourth of his life as a boy; a fifth as a youth; a third as a man; and spent thirteen years in his dotage: how old was he? Suppose we assume his age to have been 40. Then, by the given conditions, he would have spent $8\frac{2}{3}$ (and not 13) years in his dotage, and therefore

$$8\tfrac{2}{3} : 13 = 40 : \text{his actual age},$$

hence his actual age was 60. Recorde adds that he preferred to solve problems by this method since when a difficult question was proposed he could obtain the true result by taking the chance answers of "such children or idiots as happened to be in the place."

Like all his works the *Grounde of artes* is written in the form of a dialogue between master and scholar. As an illustration of the style I quote from it the introductory conversation on the advantages of the power of counting "the only thing that separateth man from beasts."

Master. If Number were so vile a thing as you did esteem it, then need it not to be used so much in mens communication. Exclude Number and answer me to this question. How many years old are you?

Scholar. Mum.

Master. How many days in a week? How many weeks in a year? What lands hath your father? How many men doth he keep? How long is it sythe you came from him to me?

Scholar. Mum.

Master. So that if Number want, you answer all by Mummes. How many miles to London?...Why, thus you may see, what rule Number beareth and that if Number be lacking, it maketh men dumb, so that to most questions, they must answer Mum.

Recorde also published in 1556 an algebra called the *Whetstone of witte.* The title, as is well known, is a play on the old name of algebra as the cossic art: the term being derived from *cosa*, a thing, which was used to denote the unknown quantity in an equation. Hence the title *cos ingenii*, the whetstone of wit. The algebra is syncopated, that is, it is written at length according to the usual rules of grammar, but symbols or contractions are used for the quantities and operations which occur most frequently. In this work Recorde shewed how the square root of an algebraical expression could be extracted—a rule which was here published for the first time.

Both these treatises were frequently republished and had a wide circulation. The latter in particular was well known, as is shewn by the allusion to it (as being studied by the usurer) in Sir Walter Scott's *Fortunes of Nigel.* To the belated traveller who wanted some literature wherewith to pass the time, the maid, says he, "returned for answer that she knew of no other books in the house than her young mistress's bible, which the owner would not lend; and her master's Whetstone of Witte by Robert Recorde." So too William Cuningham[1] in his *Cosmographicall glasse*, published in 1559, alludes to

[1] *William Cuningham* (sometimes written *Keningham*) was born in 1531 and entered at Corpus College, Cambridge, in 1548. The *Cosmographicall glasse*, is the earliest English treatise on cosmography. Cuningham also published some almanacks, but his works have no intrinsic value in the history of the mathematics. He practised as a physician in London, under the license conferred by his Cambridge degree.

Recorde's writings as standard authorities in arithmetic and algebra: in geometry he quotes Orontius and Euclid.

Besides the two books just mentioned Recorde wrote the following works on mathematical subjects. The *Pathway to knowledge*, published in 1551, on geometry and astronomy; the *Principles of geometry* also written in 1551; three works issued in 1556 on astronomy and astrology, respectively entitled the *Castle*, *Gate*, and *Treasure of knowledge*; and lastly a treatise on the sphere, and another on mensuration, both of which are undated. He also translated Euclid's *Elements*, but I do not think that this was published.

In his astronomy Recorde adopts the Copernican hypothesis. Thus in one of his dialogues he induces his scholar to assert that the "earth standeth in the middle of the world." He then goes on

> *Master.* How be it, Copernicus a man of great learning, of much experience, and of wonderful diligence in observation, hath renewed the opinion of Aristarchus of Samos, and affirmeth that the earth not only moveth circularly about his own centre, but also may be, yea and is, continually out of the precise centre 38 hundred thousand miles: but because the understanding of that controversy dependeth of profounder knowledge than in this introduction may be uttered conveniently, I will let it pass till some other time.
>
> *Scholar.* Nay sir in good faith, I desire not to hear such vain phantasies, so far against common reason, and repugnant to the consent of all the learned multitude of writers, and therefore let it pass for ever, and a day longer.
>
> *Master.* You are too young to be a good judge in so great a matter: it passeth far your learning, and theirs also that are much better learned than you, to improve his supposition by good arguments, and therefore you were best to condemn nothing that you do not well understand: but another time, as I said, I will so declare his supposition, that you shall not only wonder to hear it, but also peradventure be as earnest then to credit it, as you are now to condemn it.

This advocacy of the Copernican theory is the more remarkable as that hypothesis was only published in 1543, and was merely propounded as offering a simple explanation of the phenomena observable: Galileo was the first writer who attempted to give a proof of it. It is stated that Recorde was the earliest Englishman who accepted that theory.

Recorde's works give a clear view of the knowledge of the time and he was certainly the most eminent English mathematician of that age, but I do not think he can be credited with any original work except the rule for extracting the square root of an algebraical expression.

Another mathematician only slightly junior to Recorde was Dee, who fills no small place in the scientific and literary records of his time, and whose natural ability was of the highest order. **John Dee**[1] was born on July 13, 1527, and died in December 1608. He entered at St John's College[2] in 1542, proceeded B.A. in 1545, and was elected to a fellowship in the following year. On the foundation of Trinity College in 1546, Dee was nominated one of the original fellows, and was made assistant lecturer in Greek—a post which however he only held for a year and a half. During this time he was studying mathematics, and on going down in 1548 he gave his astronomical instruments to Trinity.

He then went on the continent. In 1549 he was teaching arithmetic and astronomy at Louvain, and in 1550 he was lecturing at Paris *in English* on Euclidean geometry. These lectures are said to have been the first gratuitous ones ever given in a European university (see p. 143). "My auditory in Rheims College" says he "was so great, and the most part elder than myself, that the mathematical schools could not hold them; for many were fain without the schools at the windows, to be auditors and spectators, as they best could help themselves thereto. I did also dictate upon every proposition besides the

[1] There are numerous biographies of Dee, which should be read in connection with his diaries. Perhaps one of the best is in Thomas Smith's *Vitae...illustrium virorum*. A bibliography of his works (seventy-nine in number) and an account of his life are given in vol. II. pp. 505–9 of the *Athenae Cantabrigienses*.

[2] Here, and hereafter when I mention a college, the reference is to the college of that name at Cambridge, unless some other university or place is expressly mentioned.

first exposition. And by the first four principal definitions representing to their eyes (which by imagination only are exactly to be conceived) a greater wonder arose among the beholders, than of my Aristophanes Scarabæus mounting up to the top of Trinity hall in Cambridge." The last allusion is to a stage trick which he had designed for the performance of a Greek comedy in the dining-hall at Trinity and which, unluckily for him, gave him the reputation of a sorcerer among those who could not see how it was effected.

In 1554 some public-spirited Oxonians, who regretted the manner in which scientific studies were there treated, offered him a stipend to lecture on mathematics at Oxford, but he declined the invitation. A year or so later we find him petitioning queen Mary to form a royal library by collecting all the dispersed libraries of the various monasteries, and it is most unfortunate that his proposal was rejected.

On the accession of Elizabeth he was taken into the royal service, and subsequently most of his time was occupied with alchemy and astrology. It is now generally admitted that in his experiments and alleged interviews with spirits he was the dupe of others and not himself a cheat. His chief work on astronomy was his report to the Government made in 1585 advocating the reform of the Julian calendar: like Recorde he adopted the Copernican hypothesis. The Government accepted his proposal but owing to the strenuous opposition of the bishops it had to be abandoned, and was not actually carried into effect till nearly two hundred years later.

During the last part of his life Dee was constantly in want, and his reputation as a sorcerer caused all men to shun him. The story of his intercourse with angels and experiments on the transmutation of metals are very amusing, but are too lengthy for me to cite here. His magic crystal and cakes are now in the British Museum.

He is described as tall, slender, and handsome, with a clear and fair complexion. In his old age he let his beard, which was then quite white, grow to an unusual length, and never appeared abroad except "in a long gown with hanging sleeves." An engraving of a portrait of him executed in his lifetime and now in my possession fully bears out this description. No doubt these peculiarities of dress added to his evil reputation as a dealer in evil spirits, but throughout his life he seems to have been constantly duped by others more skilful and less scrupulous than himself.

Among the pupils of Dee was **Thomas Digges**, who entered at Queens' College in 1546 and proceeded B.A. in 1551. Digges edited and added to the writings of his father Leonard Digges, but how much is due to each it is now impossible to say with certainty, though it is probable that the greater part is due to the son. His works in 24 volumes are mostly on the application of arithmetic and geometry to mensuration and the arts of fortification and gunnery. They are chiefly remarkable as being the earliest English books in which spherical trigonometry is used[1]. In one of them known as *Pantometria* and issued in 1571 the theodolite is described: this is the earliest known description of the instrument[2]. The derivation is from an Arabic word *alhidada* which was corrupted into *athelida* and thence into *theodelite.* Digges was muster-master of the English army, and so engrossed with political and military matters as to leave but little time for original work; but Tycho Brahe[3] and other competent observers deemed him to be one of the greatest geniuses of that time. He died in 1595.

Thomas Blundeville was resident at Cambridge about the same time as Dee and Digges—possibly he was a non-collegiate student, and if so must have been one of the last of them. In 1589 he wrote a work on the use of maps and of Ptolemy's tables. In 1594 he published his *Exercises* in six parts, containing a brief account of arithmetic, cosmography, the use of the globes, a universal map, the astrolabe, and navigation.

[1] See p. 40 of the *Companion to the Almanack for* 1837.

[2] See p. 24 of *Arithmetical books* by A. De Morgan, London, 1847.

[3] See pp. 6, 33 of *Letters on scientific subjects* edited by Halliwell, London, 1841.

The arithmetic is taken from Recorde, but to it are added trigonometrical tables (copied from Clavius) of the natural sines, tangents, and secants of all angles in the first quadrant; the difference between consecutive angles being one minute. These are worked out to seven places of decimals. This is the earliest[1] English work in which plane trigonometry is introduced.

Another famous teacher of the same period was **William Buckley**. Buckley was born at Lichfield, and educated at Eton, whence he went to King's in 1537, and proceeded B.A. in 1542. He subsequently attended the court of Edward VI., but his reputation as a successful lecturer was so considerable that about 1550 he was asked to return to King's to teach arithmetic and geometry. He has left some mnemonic rules on arithmetic which are reprinted in the second edition of Leslie's *Philosophy of arithmetic*, Edinburgh, 1820. Buckley died in 1569.

Another well known Cambridge mathematician of this time was **Sir Henry Billingsley**, who obtained a scholarship at St John's College in 1551. He is said on somewhat questionable authority to have migrated from Oxford, and to have learnt his mathematics from an old Augustinian friar named Whytehead, who continued to live in the university after the suppression of the house of his order. The latter is described as fat, dirty and uncouth, but seems to have been one of the best mathematical tutors of his time. Billingsley settled in London and ultimately became lord mayor; but he continued his interest in mathematics and was also a member of the Society of Antiquaries. He died in 1606.

Billingsley's claim to distinction is the fact that he published in 1570 the first English translation of Euclid. In preparing this he had the assistance both of Whytehead and of John Dee. In spite of their somewhat qualified disclaimers, it was formerly supposed that the credit of it was due to them

[1] See p. 42 of *Arithmetical books* by A. De Morgan, London, 1847.

rather than to him, especially as Whytehead, who had fallen into want, seems at the time when it was published to have been living in Billingsley's house. The copy of the Greek text of Theon's Euclid used by Billingsley has however been recently discovered, and is now in Princetown College, America[1]; and it would appear from this that the credit of the work is wholly due to Billingsley himself. The marginal notes are all in his writing, and contain comments on the edition of Adelhard and Campanus from the Arabic (see p. 4), and conjectural emendations which shew that his classical scholarship was of a high order.

Other contemporary mathematical writers are Hill, Bedwell, Hood, the two Harveys, and Forman. They are not of sufficient importance to require more than a word or two in passing.

Thomas Hill, who took his B.A. degree from Christ's College in 1553, wrote a work on Ptolemaic astronomy termed the *Schoole of skil*: it was published posthumously in 1599.

Thomas Bedwell entered at Trinity in 1562, was elected a scholar in the same year, proceeded B.A. in 1567, and in 1569 was admitted to a fellowship. His works deal chiefly with the applications of mathematics to civil and military engineering, and enjoyed a high and deserved reputation for practical good sense. The New River company was due to his suggestion. He died in 1595.

Thomas Hood, who entered at Trinity in 1573, proceeded B.A. in 1578, and was subsequently elected to a fellowship, was another noted mathematician of this time. In 1590 he issued a translation of Ramus's geometry, and in 1596 a translation of Urstitius's arithmetic. He also wrote on the use of the globes

[1] See a note by G. B. Halsted in vol. II. of the *American journal of mathematics*, 1878. The Greek text had been brought into Italy by refugees from Constantinople, and was first published in the form of a Latin translation by Zamberti at Venice in 1505: the original text (Theon's edition) was edited by Grynæus and published by Hervagius at Bâle in 1535.

(1590 and 1592), and the principles of surveying (1598). In 1582 a mathematical lectureship was founded in London—probably by a certain Thomas Smith of Gracechurch Street—and Hood was appointed lecturer. His books are probably transcripts of these lectures: the latter were given in the Staples chapel, and subsequently at Smith's house. Hood seems to have also practised as a physician under a license from Cambridge dated 1585.

Richard Harvey, a brother of the famous Gabriel Harvey, was a native of Saffron Walden. He entered at Pembroke in 1575, proceeded B.A. in 1578, and subsequently was elected to a fellowship. He was a noted astrologer, and threw the whole kingdom into a fever of anxiety by predicting the terrible events that would follow from the conjunction of Saturn and Jupiter, which it was known would occur on April 28, 1583. Of course nothing peculiar followed from the conjunction; but Harvey's reputation as a prophet was destroyed, and he was held up to ridicule in the tripos verses of that or the following year and hissed in the streets of the university. Thomas Nash (a somewhat prejudiced witness be it noted) in his *Pierce pennilesse*, published in London in 1592 says, "Would you in likely reason guess it were possible for any shame-swoln toad to have the spet-proof face to outlive this disgrace?" Harvey took a living, and his later writings are on theology. He died in 1599.

John Harvey, a brother of the Richard Harvey mentioned above, was also born at Saffron Walden: he entered at Queens' in 1578 and took his B.A. in 1580. He practised medicine and wrote on astrology and astronomy—the three subjects being then closely related. He died at Lynn in 1592.

Simon Forman[1], of Jesus College, born in 1552, was another mathematician of this time, who like those just mentioned combined the study of astronomy, astrology, and medicine with considerable success; though he is described, apparently with

[1] An account of Forman's life is given in the *Life of William Lilly, written by himself*, London, 1715.

good reason, as being as great a knave as ever existed. His license to practise medicine was granted by the university, and is dated 1604. He was a skilful observer and good mathematician, but I do not think he has left any writings. He died suddenly when rowing across the Thames on Sept. 12, 1611.

With the exception of Recorde, Dee, and Digges, all the above were but second-rate mathematicians; but such as they were (and they are nearly all the English mathematicians of that time of whom I know anything) it is noticeable that without a single exception they were educated at Cambridge. The prominence given to astronomy, astrology, and surveying is worthy of remark.

I come next to a group of mathematicians of considerably greater power, to whom we are indebted for important contributions to the progress of the science.

The first of these was **Edward Wright**[1], whose services to the theory of navigation can hardly be overrated. Wright was born in Norfolk, took his B.A. from Caius in 1581, and was subsequently elected to a fellowship. He seems to have had a special talent for the construction of instruments; and to instruct himself in practical navigation and see what improvements in nautical instruments were possible, he went on a voyage in 1589—special leave of absence from college being granted him for the purpose.

In the maps in use before the time of Gerard Mercator a degree whether of latitude or longitude had been represented in all cases by the same length, and the course to be pursued by a vessel was marked on the map by a straight line joining the ports of arrival and departure. Mercator had seen that this led to considerable errors, and had realized that to make this method of tracing the course of a ship at all accurate the

[1] See an article in the *Penny Cyclopaedia*, London, 1833—43; and a short note included in the article on Navigation in the ninth edition of the *Encyclopaedia Britannica*.

space assigned on the map to a degree of latitude ought gradually to increase as the latitude increased. Using this principle, he had empirically constructed some charts, which were published about 1560 or 1570. Wright set himself the problem to determine the theory on which such maps should be drawn, and succeeded in discovering the law of the scale of the maps, though his rule is strictly correct for small arcs only. The result was published by his permission in the second edition of Blundeville's *Exercises.* His reputation was so considerable that four years after his return he was ordered by queen Elizabeth to attend the Earl of Cumberland on a maritime expedition as scientific adviser.

In 1599 Wright published a work entitled *Certain errors in navigation detected and corrected*, in which he very fully explains the theory and inserts a table of meridional parts. Solar and other observations requisite for navigation are also treated at considerable length. The theoretical parts are correct, and the reasoning shews considerable geometrical power. In the course of the work he gives the declinations of thirty-two stars, explains the phenomena of the dip, parallax, and refraction, and adds a table of magnetic declinations, but he assumes the earth to be stationary. This book went through three editions. In the same year he issued a work called *The haven-finding art.* I have never seen a copy of it and I do not know how the subject is treated. In the following year he published some maps constructed on his principle. In these the northernmost point of Australia is shewn: the latitude of London is taken to be 51° 32′.

About this time Wright was elected lecturer on mathematics by the East India Company at a stipend of £50 a year. He now settled in London, and shortly afterwards was appointed mathematical tutor to prince Henry of Wales, the son of James I. He here gave another proof of his mechanical ability by constructing a sphere which enabled the spectator to forecast the motions of the solar system with such accuracy that it was possible to predict the eclipses for over seventeen thousand years in advance: it was shewn in the Tower as a curiosity as late as 1675. Wright also seems to have joined Bedwell in urging that the construction of the New River to supply London with drinking water was both feasible and desirable.

As soon as Napier's invention of logarithms was announced in 1614, Wright saw its value for all practical problems in navigation and astronomy. He at once set himself to prepare an English translation. He sent this in 1615 to Napier, who approved of it and returned it, but Wright died in the same year, before it was printed: it was issued in 1616.

Whatever might have been Wright's part in bringing logarithms into general use it was actually to Briggs, the second of the mathematicians above alluded to, that the rapid adoption of Napier's great discovery was mainly due.

Henry Briggs[1] was born near Halifax in 1556. He was educated at St John's College, took his B.A. degree in 1581, and was elected to a fellowship in 1588. He continued to reside at Cambridge, and in 1592 he was appointed examiner and lecturer in mathematics at St John's.

In 1596 the college which *Sir Thomas Gresham*[2] had directed to be built was opened. Gresham, who was born in 1513 and died in 1579, had been educated at Gonville Hall, and had apparently made some kind of promise to build the college at Cambridge to encourage research, so that his final determination to locate it in London was received with great disappointment in the university. The college was endowed for the study of the seven liberal sciences; namely, divinity, astronomy, geometry, music, law, physic, and rhetoric.

Briggs was appointed to the chair of geometry. He seems at first to have occupied his leisure in London by researches on

[1] See the *Lives of the professors of Gresham College* by J. Ward, London, 1740. A full list of Briggs's works is given in the *Dictionary of national biography*.

[2] See the *Life and times of Sir Thomas Gresham*, published anonymously but I believe written by J. W. Burgon, London, 1845.

magnetism and eclipses. Almost alone among his contemporaries he declared that astrology was at best a delusion even if it were not, as was too frequently the case, a mere cloak for knavery. In 1610 he published *Tables for the improvement of navigation*, and in 1616 a *Description of a table to find the part proportional devised by Edw. Wright.*

In 1614 Briggs received a copy of Napier's work on logarithms, which was published in that year. He at once realized the value of the discovery for facilitating all practical computations, and the rapidity with which logarithms came into general use was largely due to his advocacy. The base to which the logarithms were at first calculated was very inconvenient, and Briggs accordingly visited Napier in 1616, and urged the change to a decimal base, which was recognized by Napier as an improvement. Briggs at once set to work to carry this suggestion into effect, and in 1617 brought out a table of logarithms of the numbers from 1 to 1000 calculated to fourteen places of decimals. He subsequently (in 1624) published tables of the logarithms of additional numbers and of various trigonometrical functions. The calculation of some 20,000 logarithms which had been left out by Briggs in his tables of 1624 was performed by Vlacq and published in 1628. The *Arithmetica logarithmica* of Briggs and Vlacq are substantially the same as the existing tables: parts have been recalculated, but no tables of an equal range and fulness entirely founded on fresh computations have since been published. These tables were supplemented by Briggs's *Trigonometrica Britannica* which was published posthumously in 1633.

The introduction of the decimal notation was also (in my opinion) due to Briggs. Stevinus in 1585, and Napier in his essay on rods in 1617, had previously used a somewhat similar notation, but they only employed it as a concise way of stating results, and made no use of it as an operative form. The notation occurs however in the tables published by Briggs in 1617, and was adopted by him in all his works, and though it is difficult to speak with absolute certainty I have myself but little doubt that he there employed the symbol as an operative form. In Napier's posthumous *Constructio* published in 1619 it is defined and used systematically as an operative form, and as this work was written after consultation with Briggs, and was probably revised by him before it was issued, I think it confirms the view that the invention was due to Briggs and was communicated by him to Napier. At any rate its use as an operative form was not known to Napier in 1617. Napier wrote the point in the form now adopted, but Briggs underlined the decimal figures, and would have printed a number such as 25·379 in the form $25\underline{379}$. Later writers added another line and wrote it $25\underline{|379}$; nor was it till the beginning of the eighteenth century that the notation now current was generally employed.

Shortly after bringing out the first of his logarithmic tables, Briggs moved to Oxford. For more than two centuries—possibly from the time of Bradwardine—Merton had been the one college in that university where instruction in mathematics had been systematically given. When Sir Henry Savile (born in 1549 and died in 1622) became warden of Merton he seems to have felt that the practical abandonment of science to Cambridge was a reproach on the ancient and immensely more wealthy university of Oxford. Accordingly about 1570 he began to give lectures on Greek geometry, which, contrary to the almost universal practice of that age, he opened free to all members of the university. These lectures were published at Oxford in 1621. He never however succeeded in taking his class beyond the eighth proposition of the first book of Euclid. "Exolvi," says he, "per Dei gratiam, domini auditores; promissum; liberavi fidem meam; explicavi pro meo modulo, definitiones, petitiones, communes sententias, et octo priores propositiones Elementorum Euclidis. Hic, annis fessus, cyclos artemque repono."

In spite of this discouraging result Savile hoped to make the study a permanent one, and in 1619 he founded two chairs, one of geometry and one of astronomy. The former he offered to Briggs, who thus has the singular distinction of holding in succession the two earliest chairs of mathematics that were founded in England. Briggs continued to hold this post until his death on Jan. 26, 1630.

Among Briggs's contemporaries at Cambridge was Oughtred, who systematized elementary arithmetic, algebra, and trigonometry. **William Oughtred**[1] was born at Eton on March 5, 1574. He was educated at Eton, and thence in 1592 went to King's College. While an undergraduate he wrote an essay on geometrical dialling. He took his B.A. degree in 1596, was admitted to a fellowship in the ordinary course, and lectured for a few years; but on taking orders in 1603 he felt it his duty to devote his time wholly to parochial work.

Although living in a country vicarage he kept up his interest in mathematics. Equally with Briggs he received one of the earliest copies of Napier's *Canon* on logarithms, and was at once impressed with the great value of the discovery. Somewhat later in life he wrote two or three works. He always gave gratuitous instruction to any who came to him, provided they would learn to "write a decent hand." He complained bitterly of the penury of his wife who always took away his candle after supper "whereby many a good motion was lost and many a problem unsolved"; and one of his pupils who secretly gave him a box of candles earned his warmest esteem. He is described as a little man, with black hair, black eyes, and a great deal of spirit. Like nearly all the mathematicians of the time he was somewhat of an astrologer and alchemist. He died at his vicarage of Albury in Surrey on June 30, 1660.

His *Clavis mathematica* published in 1631 is a good systematic text-book on algebra and arithmetic, and it contains practically all that was then known on the subject. In this work he introduced the symbol × for multiplication, and the

[1] See *Letters...and lives of eminent men* by J. Aubrey, 2 vols., London, 1813. A complete edition of Oughtred's works was published at Oxford in 1677.

symbol :: in proportion. Previously to his time a proportion such ac $a : b = c : d$ was written as $a - b - c - d$, but he denoted it by $a . b :: c . d$. Wallis says that some found fault with the book on account of the style, but that they only displayed their own incompetence, for Oughtred's "words be always full but not redundant." Pell makes a somewhat similar remark.

A work on sun and other dials published in 1636 shews considerable geometrical power, and explains how various astronomical problems can be resolved by the use of dials. He also wrote a treatise on trigonometry published in 1657 which is one of the earliest works containing abbreviations for *sine*, *cosine*, &c. This was really an important advance, but the book was neglected and soon forgotten, and it was not until Euler reintroduced contractions for the trigonometrical functions that they were generally adopted.

The following list comprises all his works with which I am acquainted. The *Clavis*, first edition 1631; second edition with an appendix on numerical equations 1648; third edition greatly enlarged, 1652. *The circle of proportion*, 1632; second edition 1660. *The double horizontal dial*, 1636; second edition 1652. *Sun-dials by geometry*, 1647. *The horological ring*, 1653. *Solution of all spherical triangles*, 1657. *Trigonometry*, 1657. *Canones sinuum etc.*, 1657. And lastly a posthumous work entitled *Opuscula mathematica hactenus inedita*, issued in 1677.

Just as Briggs was the most famous English geometrician of that time, so to his contemporaries Oughtred was probably the most celebrated exponent of algorism. Thus in some doggrel verses in the *Lux mercatoria* by Noah Bridges, London, 1661, we read that a merchant

"may fetch home the Indies, and not know
what Napier could or what Oughtred can do."

Another mathematician of this time, who was almost as well known as Briggs and Oughtred, was *Thomas Harriot* who was born in 1560, and died on July 2, 1621. He was not educated at either university, and his chief work the *Artis analyticae praxis* was not printed till 1631. It is incomparably the best work on algebra and the theory of equations which had then been published. I mention it here since it became a recognized text-book on the subject, and for at least a century the more advanced Cambridge undergraduates, including Newton, Whiston, Cotes, Smith, and others, learnt most of their algebra thereout. We may say roughly that henceforth elementary arithmetic, algebra, and trigonometry were treated in a manner which is not substantially different from that now in use; and that the subsequent improvements introduced are additions to the subjects as then known, and not a re-arrangement of them on new foundations.

The work of most of those considered in this chapter—which we may take as comprised between the years 1535 and 1630—is manifestly characterized by the feeling that mathematics should be studied for the sake of its practical applications to astronomy (including astrology therein), navigation, mensuration, and surveying; but it was tacitly assumed that even in these subjects its uses were limited, and that a knowledge of it was in no way necessary to those who applied the rules deduced therefrom, while it was generally held that its study did not form any part of a liberal education.

CHAPTER III.

THE COMMENCEMENT OF MODERN MATHEMATICS.

In the last chapter I was able to trace a continuous succession of mathematicians resident at Cambridge to the end of the sixteenth century. The period of the next thirty years is almost a blank in the history of science at the university, but its close is marked by the publication of some of the more important works of Briggs, Oughtred, and Harriot. We come then to the names of Horrox and Seth Ward, both of whom were well-known astronomers; to Pell, who was later in intimate relations with Newton; and lastly to Wallis and to Barrow, who were the first Englishmen to treat mathematics as a science rather than as an art, and who may be said to have introduced the methods of modern mathematics into Britain. It curiously happened that in the absence of any endowments for mathematics at Cambridge both Ward and Wallis were elected to professorships at Oxford, and by their energy and tact created the Oxford mathematical school of the latter half of the seventeenth century.

The middle of the seventeenth century marks the beginning of a new era in mathematics. The invention of analytical geometry and the calculus completely revolutionized the development of the subject, and have proved the most powerful instruments of modern progress. Descartes's geometry was published in 1637 and Cavalieri's method of indivisibles, which is equivalent to integration regarded as a means of summing series, was introduced a year or so later. The works of both these writers were very obscure, but they had a wide circulation, and we may say that by about 1660 the methods used by them were known to the leading mathematicians of Europe. This was largely due to the writings of Wallis. Barrow occupies a position midway between the old and the new schools. He was acquainted with the elements of the new methods, but either by choice or through inability to recognize their powers he generally adhered to the classical methods. It was to him that Newton was indebted for most of his instruction in mathematics; he certainly impressed his contemporaries as a man of great genius, and he came very near to the invention of the differential calculus.

The infinitesimal calculus was invented by Newton in 1666, and was among the earliest of those discoveries and investigations which have raised him to the unique position which he occupies in the history of mathematics. The calculus was not however brought into general use till the beginning of the eighteenth century. The discoveries of Newton materially affected the whole subsequent history of mathematics, and at Cambridge they led to a complete rearrangement of the system of education. It will therefore be convenient to defer the consideration of his life and works to the next chapter.

The chief distinction between the classical geometry and the method of exhaustions on the one hand, and the new methods introduced by Descartes, Cavalieri, and Newton on the other is that the former required a special procedure for every particular problem attacked, while in the latter a general rule is applicable to all problems of the same kind. The validity of this process is proved once for all, and it is no longer requisite to invent some special process for every separate function, curve, or surface.

Another cause which makes it desirable to take this time as the commencement of a new chapter is the change in the character of the scholastic exercises in the university which then first began to be noticeable. The disturbances produced by the civil wars in the middle of the seventeenth century did not affect Cambridge so severely as Oxford, but still they produced considerable disorder, and thenceforward the regulations of the statutes about exercises in the schools seem to have been frequently disregarded. The Elizabethan statutes had directed that logic should form the basis of a university education, and that it should be followed by a study of Aristotelian philosophy. The logic that was read at Cambridge was that of Ramus. This was purely negative and destructive, and formed an admirable preparation for the Baconian and Cartesian systems of philosophy. The latter were about this time adopted in lieu of a study of Aristotle, and they provided the usual subject for discussions in the schools for the remainder of the seventeenth century, until in their turn they were displaced by the philosophy of Newton and of Locke[1].

I shall commence by a very brief summary of the views of Horrox and Seth Ward, and shall then enumerate some other contemporary astronomers of less eminence. I shall next describe the writings of Pell, Wallis, and Barrow; and it will be convenient to add references to a few other mathematicians the general character of whose works is pre-newtonian.

Jeremiah Horrox[2]—sometimes written Horrocks—was born near Liverpool in 1619; he entered at Emmanuel College in 1633, but probably went down without taking a degree in 1635 or 1636; he died in 1641. From boyhood he had resolved to make himself an astronomer. No astronomy seems then to have been taught at Cambridge, and Horrox says that he had chiefly to rely on reading books by himself. He had but small means; and desiring that his library should contain only the best works on the subject he took a great deal of

[1] See p. 69 of *On the Statutes* by G. Peacock, London, 1841.

[2] See his life by A. B. Whatton, second edition, London, 1875. The works of Horrox were collected by Wallis and published at London in 1672.

trouble in selecting them. The list he drew up, written at the end of his copy of Lansberg's tables, is now in the library of Trinity and sufficiently instructive to deserve quotation.

Albategnius.
Alfraganus.
J. Capitolinus.
Clavii Apolog. Cal. Rom.
Clavii Comm. in Sacroboscum.
Copernici Revolutiones.
Cleomedes.
Julius Firmicus.
Gassendi Exerc. Epist. in Phil. Fluddanam.
Gemmæ Frisii Radius Astronomicus.
Cornelii Gemmæ Cosmocritice.
Herodoti Historia.
J. Kepleri Astron. Optica.
——— Epit. Astron. Copern.
——— Comm. de Motu Martis.
J. Kepleri Tabulæ Rudolphinæ.
Lansbergii Progymn. de Motu Solis.
Longomontani Astron. Danica.
Magini Secunda Mobilia.
Mercatoris Chronologia.
Plinii Hist. Naturalis.
Ptolemæi Magnum Opus.
Regiomontani Epitome.
——— Torquetum.
——— Observata.
Rheinoldi Tab. Prutenicæ.
—— Comm. in Theor. Purbachii.
Theonis Comm. in Ptolom.
Tyc. Brahæi Progymnasmata.
——— Epist. Astron.
Waltheri Observata.

This list probably represents the most advanced astronomical reading of the Cambridge of that time.

In spite of his early death Horrox did more to improve the lunar theory than any Englishman before Newton; and in particular he was the first to shew that the lunar orbit might be exactly represented by an ellipse, provided an oscillatory motion were given to the apse line and the eccentricity made to vary. This result was deduced from the law of gravitation by Newton in the thirty-fifth proposition of the third book of the *Principia*. Horrox was also the first observer who noted that Venus could be seen on the face of the sun: the observation was made on Nov. 24, 1639, and an account of it was printed by Hevelius at Danzig in 1662.

Seth Ward[1] was born in Hertfordshire in 1617, took his B.A. from Sidney Sussex College in 1637 at the same time as Wallis, and was subsequently elected a fellow. In his

[1] See his life by Walter Pope, London, 1697; and *Letters......and lives of eminent men* by J. Aubrey, 2 vols., London, 1813.

dispute with the prævaricator in 1640, he was publicly rebuked for the freedom of his language and his supplicat for the M.A. degree rejected, but the censure seems to have been undeserved and was withdrawn. He was celebrated for his knowledge of mathematics and especially of astronomy; and he was also a man of considerable readiness and presence. While residing at Cambridge he taught, and one of his pupils says that he "brought mathematical learning into vogue in the university...where he lectured his pupils in Master Oughtred's *Clavis*."

He was expelled from his fellowship by the parliamentary party for refusing to subscribe the league and covenant. On this Oughtred invited him to his vicarage, where he could pursue his mathematical studies without interruption. His companion on this visit was a certain Charles Scarborough, a fellow of Caius and described as a teacher of the mathematics at Cambridge, of whom I know nothing more.

In 1649 Ward was appointed to the Savilian chair of astronomy at Oxford and, like Wallis who was appointed at the same time, consented, with some hesitation, to take the oath of allegiance to the commonwealth. The two mathematicians who had been together at Cambridge exerted themselves with considerable success to revive the study of mathematics at Oxford; and they both took a leading part in the meetings of the philosophers, from which the Royal Society ultimately developed. Ward proceeded to a divinity degree in 1654, and subsequently held various ecclesiastical offices, including the bishoprics of Exeter and Salisbury. He died in January, 1689.

Aubrey describes him as singularly handsome, though perhaps somewhat too fond of athletics, at which he was very proficient. Courteous, rich, generous, with great natural abilities, and wonderful tact, he managed to make all men trust his honour and desire his friendship—a somewhat astonishing feat in those troubled times.

He wrote a text-book on trigonometry published at Oxford

in 1654, but he is best known for his works on astronomy. These are two in number, namely, one on comets and the hypothesis of Bulialdus published at Oxford in 1653; and the other on the planetary orbits published in London in 1656. The hypothesis of Bulialdus, which Ward substantially adopted, is that for every planetary orbit there is a point (called the upper focus) on the axis of the right cone of which the orbit is a section such that the radii vectores thence drawn to the planet move with a uniform motion: the idea being the same as that held by the Greeks, namely, that the motion of a celestial body must be perfect and therefore must be uniform.

Other astronomers of the same time were Samuel Foster, Laurence Rooke, Nicholas Culpepper, and Gilbert Clerke. I add a few notes on them.

Samuel Foster[1], of Emmanuel College, who was born in Northamptonshire, took his B.A. in 1619, and in 1636 was appointed Gresham professor of astronomy, but was shortly expelled for refusing to kneel when at the communion table: he was however reappointed in 1641, and held the chair till his death, which took place in 1652. He wrote several works, of which a list is given on pp. 86–87 of Ward's *Lives*: most of them are on astronomical instruments, but one volume contains some interesting essays on various problems in Greek geometry. Foster took a prominent part in the meetings of the so-called "indivisible college" during the year 1645, from which the Royal Society ultimately sprang.

Foster was succeeded in his chair at Gresham College by Rooke. **Laurence Rooke**[1], who was born in Kent in 1623, took his B.A. in 1643 from King's College. He lectured at Cambridge on Oughtred's *Clavis* for some time after his degree. Like Foster he took a leading part in the meetings of the indivisible college: being a man of considerable property he assisted the society in several ways, and in 1650 he moved to Oxford with

[1] See the *Lives of the professors of Gresham College* by J. Ward, London, 1740.

most of the other members. In 1652 he was appointed professor of astronomy at Gresham College, and in 1657 he exchanged it for the chair of geometry, which he held till his death in 1662. His lectures were given on the sixth chapter of Oughtred's *Clavis*, which enables us to form an idea of the extent of mathematics then usually known. A list of his writings is given in Ward: most of them are concerned with various practical questions in astronomy.

Nicholas Culpepper, of Queens', who was born in London on Oct. 18, 1616, entered at Cambridge in 1634 and died on Jan. 10, 1653, was a noted astrologer of the time. He used his knowledge of astronomy to justify various medical remedies employed by him, which though they savoured of heresy to the orthodox practitioner of that day, seem to have been fairly successful. It is doubtful whether he was a quack or an unpopular astronomer. I suspect he has a better claim to the former title than the latter one, but I give him the benefit of the doubt. His works, edited by G. A. Gordon, were published in four volumes in London in 1802.

Gilbert Clerke, a fellow of Sidney College, was born at Uppingham in 1626, and graduated B.A. in 1645. He lectured for a few years at Cambridge, but in 1655 was forced to quit the university by the Cromwellian party. He had a small property in Norfolk and lived there till his death. His chief mathematical works were the *De plenitudine mundi*, published in 1660, in which he defended Descartes from the criticisms of Bacon and Seth Ward; an account of some experiments analogous to those of Torricelli, published in 1662; a commentary on Oughtred's *Clavis*, published in 1682; and a description of the "spot-dial," published in 1687. He was a friend of Cumberland and of Whiston. He died towards the end of the seventeenth century.

The three mathematicians to be next mentioned—Pell, Wallis, and Barrow—were men of much greater mark, and in their writings we begin to find mathematics treated as a science.

John Pell[1] was born in Sussex on March 1, 1610: he entered at Trinity at the unusually early age of thirteen, and proceeded to his degrees in regular course, commencing M.A. in 1630. After taking his degree he continued the study of mathematics, and his reputation was so considerable that in 1639 he was asked to stand for the mathematical chair then vacant at the university of Amsterdam; but he does not seem to have gone there till 1643. In 1646 he moved, at the request of the prince of Orange, to the college which the latter had just founded at Breda. In 1654 he entered the English diplomatic service, and in 1661 took orders and became private chaplain to the archbishop of Canterbury. He still however continued the study of philosophy and mathematics to the no small detriment of his private affairs. It was to him that Newton about this time explained his invention of fluxions. He died in straitened circumstances in London on Dec. 10, 1685.

He was especially celebrated among his contemporaries for his lectures on the algebra of Diophantus and the geometry of Apollonius, of which authors he had made a special study. He had prepared these lectures for the press, but their publication was abandoned at the request of one of his Dutch colleagues. In 1668 he issued in London a new edition of Branker's translation from the Dutch of Rhonius's algebra, with the addition of considerable new matter: in this work the symbol ÷ for division was first employed. In 1672 he published at London a table of all square numbers less than 10^8. These were his chief works, but he also wrote an immense number of

[1] See the *Penny Cyclopaedia*, London, 1833—43. The custom which prevailed amongst the more wealthy classes of obtaining as soon as possible the horoscope of a child enables us to fix the date of birth with far greater accuracy than might have been expected by those unacquainted with the habits of the time. Pell for example was born at 1.21 p.m. on the day above mentioned.

pamphlets and letters on various scientific questions then debated: those now extant fill nearly fifty folio volumes, and a competent review of them would probably throw considerable light on the scientific history of the seventeenth century, and possibly on the state of university education in the first half of that century.

The following are the titles and dates of his published writings. *On the quadrant*, 2 vols., 1630. *Modus supputandi ephemerides*, 1630. *On logarithms*, 1631. *Astronomical history*, 1633. *Foreknower of eclipses*, 1633. *Deduction of astronomical tables from Lansberg's tables*, 1634. *On the magnetic needle*, 1635. *On Easter*, 1644. *An idea of mathematics*, 1650. *Branker's translation of Rhonius's algebra*, 1668. *A table of square numbers*, 1672.

The next and by far the most distinguished of the mathematicians of this time is Wallis. **John Wallis**[1] was born at Ashford on Nov. 22, 1616. When fifteen years old he happened to see a book of arithmetic in the hands of his brother; struck with curiosity at the odd signs and symbols in it he borrowed the book, and in a fortnight had mastered the subject. It was intended that he should be a doctor, and he was sent to Emmanuel College, the chief centre of the academical puritans. He took his B.A. in 1637; and for that kept one of his acts, on the doctrine of the circulation of the blood—this was the first occasion on which this theory was publicly maintained in a disputation.

His interests however centred on mathematics. Writing in 1635 he gives an account of his undergraduate training. He says that he had first to learn logic, then ethics, physics, and metaphysics, and lastly (what was worse) had to consult the schoolmen on these subjects. Mathematics, he goes on, were "scarce looked upon as Academical studies, but rather

[1] See the *Biographia Britannica*, first edition, London, 1747—66, and the *Histoire des sciences mathématiques* by M. Marie, Paris, 1833—88. Wallis's mathematical works were published in three volumes at Oxford, 1693—99.

Mechanical...And among more than two hundred students (at that time) in our college, I do not know of any two (perhaps not any) who had more of Mathematicks than I, (if so much) which was then but little; and but very few, in that whole university. For the study of Mathematicks was at that time more cultivated in London than in the universities." This passage has been quoted as shewing that no attention was paid to mathematics at that time. I do not think that the facts justify such a conclusion; at any rate Wallis, whether by his own efforts or not, acquired sufficient mathematics at Cambridge to be ranked as the equal of mathematicians such as Descartes, Pascal, and Fermat.

Wallis was elected to a fellowship at Queens', commenced M.A. in 1640, and subsequently took orders, but on the whole adhered to the puritan party to whom he rendered great assistance in deciphering the royalist despatches. He however joined the moderate presbyterians in signing the remonstrance against the execution of Charles I., by which he incurred the lasting hostility of the Independents—a fact which when he subsequently lived at Oxford did something to diminish his unpopularity as a mathematician and a schismatic.

There was then no professorship in mathematics and no opening for a mathematician to a career at Cambridge; and so Wallis reluctantly left the university. In 1649 he was appointed to the Savilian chair of geometry at Oxford, where he lived until his death on Oct. 28, 1703. It was there that all his mathematical works were published. Besides those he wrote on theology, logic, and philosophy; and was the first to devise a system for teaching deaf-mutes. I do not think it necessary to mention his smaller pamphlets, a full list of which would occupy some four or five pages: but I add a few notes on his more important mathematical writings.

The most notable of these was his *Arithmetica infinitorum*, which was published in 1656. It is prefaced by a short tract on conic sections which was subsequently expanded into a separate treatise. He then established the law of indices, and shewed that x^{-n} stood for the reciprocal of x^n and that $x^{p/q}$ stood for the qth root of x^p. He next proceeded to find by the method of indivisibles the area enclosed between the curve $y=x^m$, the axis of x, and any ordinate $x=h$; and he proved that this was to the parallelogram on the same base and of the same altitude in the ratio $1:m+1$. He apparently assumed that the same result would also be true for the curve $y=ax^m$, where a is any constant. In this result m may be any number positive or negative, and he considered in particular the case of the parabola in which $m=2$, and that of the hyperbola in which $m=-1$: in the latter case his interpretation of the result is incorrect. He then shewed that similar results might be written down for any curve of the form $y=\Sigma ax^m$; so that if the ordinate y of a curve could be expanded in powers of the abscissa x, its quadrature could be determined. Thus he said that if the equation of a curve was $y=x^0+x^1+x^2+\dots$ its area would be $x+\frac{1}{2}x^2+\frac{1}{3}x^3+\dots$. He then applied this to the quadrature of the curves $y=(1-x^2)^0$, $y=(1-x^2)^1$, $y=(1-x^2)^2$, $y=(1-x^2)^3$, &c. taken between the limits $x=0$ and $x=1$: and shewed that the areas are respectively

$$1,\ \tfrac{2}{3},\ \tfrac{8}{15},\ \tfrac{16}{35},\ \&c.$$

He next considered curves of the form $y=x^{\frac{1}{m}}$ and established the theorem that the area bounded by the curve, the axis of x, and the ordinate $x=1$, is to the area of the rectangle on the same base and of the same altitude as $m:m+1$. This is equivalent to finding the value of $\int_0^1 x^{\frac{1}{m}}dx$. He illustrated this by the parabola in which $m=2$. He stated but did not prove the corresponding result for a curve of the form $y=x^{p/q}$.

Wallis shewed great ingenuity in reducing curves to the forms given above, but as he was unacquainted with the binomial theorem he could not effect the quadrature of the circle, whose equation is $y=(1-x^2)^{\frac{1}{2}}$, since he was unable to expand this in powers of x. He laid down however the principle of interpolation. He argued that as the ordinate of the circle is the geometrical mean between the ordinates of the curves $y=(1-x^2)^0$ and $y=(1-x^2)^1$, so as an approximation its area might be taken as the geometrical mean between 1 and $\frac{2}{3}$. This is equivalent to taking $4\sqrt{\frac{2}{3}}$ or 3·26... as the value of π. But, he continued, we have in fact a series 1, $\frac{2}{3}$, $\frac{8}{15}$, $\frac{16}{35}$, ... and thus the term interpolated between 1 and $\frac{2}{3}$ ought to be so chosen as to obey the law of this series. This by an elaborate method leads to a value for the interpolated term which is equivalent to making

$$\pi = 2\,\frac{2 \,.\, 2 \,.\, 4 \,.\, 4 \,.\, 6 \,.\, 6 \,.\, 8 \,.\, 8 \,\ldots}{1 \,.\, 3 \,.\, 3 \,.\, 5 \,.\, 5 \,.\, 7 \,.\, 7 \,.\, 9 \,\ldots}.$$

The subsequent mathematicians of the seventeenth century constantly used interpolation to obtain results which we should attempt to obtain by direct algebraic analysis.

The *Arithmetica infinitorum* was followed in 1656 by a tract on the angle of contact; in 1657 by the *Mathesis universalis*; in 1658 by a correspondence with Fermat; and by a long controversy with Hobbes on the quadrature of the circle.

In 1659 Wallis published a tract on cycloids in which incidentally he explained how the principles laid down in his *Arithmetica infinitorum* could be applied to the rectification of algebraic curves: and in the following year one of his pupils, by name William Neil, applied the rule to rectify the semicubical parabola $x^3 = ay^2$. This was the first case in which the length of a curved line was determined by mathematics, and as all attempts to rectify the ellipse and hyperbola had (necessarily) been ineffectual, it had previously been generally supposed that no curves could be rectified.

In 1665 Wallis published the first systematic treatise on *Analytical conic sections*. Analytical geometry was invented by Descartes, and the first exposition of it was given in 1637: that exposition was both difficult and obscure, and to most of his contemporaries, to whom the method was new, it must have been incomprehensible. Wallis made the method intelligible to all mathematicians. This is the first book in which these curves are considered and defined as curves of the second degree and not as sections of a cone.

In 1668 he laid down the principles for determining the effects of the collision of imperfectly elastic bodies. This was followed in 1669 by a work on statics (centres of gravity), and in 1670 by one on dynamics: these provide a convenient synopsis of what was then known on the subject.

In 1686 Wallis published an *Algebra*, preceded by a historical account of the development of the subject which contains a great deal of valuable information and in which he seems to have been scrupulously fair in trying to give the credit of different discoveries to their true originators. This algebra is noteworthy as containing the first systematic use of formulæ. A given magnitude is here represented by the numerical ratio which it bears to the unit of the same kind of magnitude: thus when Wallis wanted to compare two lengths he regarded each as containing so many units of length. This will perhaps be made clearer if I say that the relation between the space described in any time by a particle moving with a uniform velocity would be denoted by Wallis by the formula $s = vt$, where s is the number representing the ratio of the space described to the unit of length; while previous writers would have denoted the same relation by stating what is equivalent to the proposition $s_1 : s_2 = v_1t_1 : v_2t_2$: (see e.g. Newton's *Principia*, bk. I. sect. I., lemma 10 or 11). It is curious to note that Wallis rejected as absurd and inconceivable the now usual idea of a negative number as being less than nothing, but accepted the view that it is something greater than infinity. The latter opinion may be right and consistent with the former, but it is hardly a more simple one.

I have already stated that the writings of Wallis published between 1655 and 1665 revealed and explained to all students the principles of those new methods which distinguish modern from classical mathematics. His reputation has been somewhat overshadowed by that of Newton, but his work was absolutely first class in quality. Under his influence a brilliant mathematical school arose at Oxford. In particular I may mention Wren, Hooke, and Halley as among the most eminent of his pupils. But the movement was shortlived, and there were no successors of equal ability to take up their work.

I come next to Barrow, the earliest occupant of the Lucasian chair at Cambridge. **Isaac Barrow**[1] was born in London in 1630 and died at Cambridge in 1677. He went to school first at Charterhouse (where he was so troublesome that his father was heard to pray that if it pleased God to take any of his children he could best spare Isaac), and subsequently to Felstead. He entered at Trinity in 1644, took his bachelor's degree in 1648, and was elected to a fellowship in 1649, at the same time as his friend John Ray, the famous botanist. He then resided for a few years in college, where he took pupils. It was for two of them that he translated the whole of Euclid's *Elements:* this remained a standard English text-book for half a century (see p. 84). In 1655 he was driven out of the country by the persecution of the Independents. A few months before, in 1654, he delivered a speech from which I quote the following passage as it throws some light on the study of mathematics at Cambridge at that time.

Nempe Euclidis, Archimedis, Ptolemæi, Diophanti horrida olim nomina jam multi e vobis non tremulis auribus excipiunt. Quid memorem jam vos didicisse, arithmeticæ ope, facili et instantaneâ operâ vel arenarum enormes numeros accurate computare, etiamsi illæ non tantum, ut fit, maris littoribus adjacerent, sed etiam ingenti cumulo quaquaversus ad primum mobile et extremas Mundi oras pertingerent: rem vulgo miram et arduam creditu, at vobis effectu facilem et expeditam? Quid, quando Geometriæ subsidio, non solum terrarum longe dissitos tractus, sed et patentissimas Cœli regiones emetiri nostis, interim ipsi quietem agentes, nec loco omnino cedentes, ad prælongas regulas catenasve immenso spatio applicandas? Quid referam alios, sublimibus alis ingenii

[1] A very appreciative account of the academical life and surroundings of Barrow by W. Whewell is prefixed to vol. IX. of A. Napier's edition of *Barrow's works*, Cambridge 1859. Another account of his life is given in the *Lives of the professors of Gresham College* by J. Ward, London, 1740. Barrow's lectures were edited by W. Whewell, Cambridge, 1860.

supremum æthera conscendentes, astrorum vestigiis presse inhærere, paratos districtim dicere, quam magna, et quam alta sunt; quantum sui circuli, et quo tempore conficiant, et qualem orbitam describant, quasi non cum nobis in hisce terris, sed cum superis in palatio Dei omnipotentis ætatem transigerent? Sane de horribili monstro, quod Algebram nuncupant, domito et profligato multi e vobis fortes viri triumpharunt: permulti ausi sunt Opticem directo obtutu inspicere; alii subtiliorem Dioptrices et utilissimam doctrinam irrefracto ingenii radio penetrare. Nec vobis hodie adeo mirabile est, Catoptrices principia et leges Mechanicæ non ignorantibus, quo artificio magnus Archimedes Romanas naves comburere potuit, nec a tot seculis immobilem Vestam quomodo stantem terram concutere potuisset.

Barrow returned to England in 1659, and in the following year he was ordained and appointed to the professorship of Greek at Cambridge; in 1662 he was also made professor of geometry at Gresham College. In the same year a chair of mathematics was founded at Cambridge under the will of Henry Lucas, of St John's College, one of the members of parliament for the university, and Barrow was selected as the first occupant[1] of it.

His lectures, delivered in 1664, 1665, and 1666, were published in 1683 under the title *Lectiones mathematicae*: these are mostly on the metaphysical basis for mathematical truths. His lectures for 1667 were published in the same year, and suggest the analysis by which Archimedes was led to his chief results.

In 1669 he issued his *Lectiones opticae et geometricae*, which is his most important work. In the part on optics many

[1] The successive professors were as follows. From 1664 to 1669, Isaac Barrow of Trinity; from 1669 to 1702, Sir Isaac Newton of Trinity (see chapter IV.); from 1702 to 1711, William Whiston of Clare (see p. 83); from 1711 to 1739, Nicholas Saunderson of Christ's (see p. 86); from 1739 to 1760, John Colson of Emmanuel (see p. 100); from 1760 to 1798, Edward Waring of Magdalene (see p. 101); from 1798 to 1820, Isaac Milner of Queens' (see p. 102); from 1820 to 1822, Robert Woodhouse of Caius (see p. 118); from 1822 to 1826, Thomas Turton of St Catharine's (see p. 118 *n.*); from 1826 to 1828, Sir George Biddell Airy of Trinity (see p. 132); from 1828 to 1839, Charles Babbage of Trinity (see p. 125); from 1839 to 1849, Joshua King of Queens' (see p. 132); who was succeeded by the present professor, G. G. Stokes of Pembroke.

problems connected with the reflexion and refraction of light are treated with great ingenuity. The geometrical focus of a point seen by reflexion or refraction is defined; and it is explained that the image of an object is the locus of the geometrical foci of every point on it. A few of the easier properties of thin lenses are also worked out, and the Cartesian explanation of the rainbow is simplified. The geometrical lectures contain some new ways of determining the areas and tangents of curves. The latter is solved by a rule exactly analogous to the procedure of the differential calculus, except that a separate determination of what is really a differential coefficient had to be made for every curve to which it was applied. Thus he took the equation of the curve between the coordinates x and y[1], gave x a very small decrement e and found the consequent decrement of y, which he represented by a. The limit of the ratio a/e when the squares of a and e were neglected was defined as the angular coefficient of the tangent at the point, and completely determined the tangent there.

Barrow's lectures failed to attract any considerable audiences, and on that account he felt conscientious scruples about retaining his chair. Accordingly in 1669 he resigned it to his pupil Newton, whose abilities he had been one of the earliest to detect and encourage. For the remainder of his life Barrow devoted most of his time to the study of divinity. In 1675 he issued an edition in one volume of the works of Archimedes, the first four books of the *Conics* of Apollonius, and the treatise of Theodosius on the sphere. He was appointed master of Trinity College in 1672, and died in 1677.

He is described as "low in stature, lean, and of a pale complexion," slovenly in his dress, and an inveterate smoker. He was noted for his strength and courage, and once when travelling in the East he saved the ship by his own prowess from capture by pirates. A ready and caustic wit made him a

[1] He actually denotes the coordinates by p and m, but I alter them to agree with the modern practice. For further details of his procedure see pp. 269—70 of my *History of mathematics*, London, 1888.

favorite of Charles II., and induced the courtiers to respect even if they did not appreciate him. He wrote with a sustained and somewhat stately eloquence, and with his blameless life and scrupulous conscientiousness was one of the most impressive characters of the time.

Before proceeding to Newton, who succeeded Barrow in the Lucasian chair and whose writings profoundly modified the subsequent development not only of the Cambridge school of mathematics but of the university system of education, I will mention three mathematicians of no great note whose works or teaching belong to the pre-newtonian age. These are Dacres, Tooke, and Morland.

Arthur Dacres, a fellow of Magdalene, was born in 1624, and proceeded B.A. in 1645. He then studied medicine and settled in London, where he occupied a leading position. He however kept up his acquaintance with mathematics, and in 1664 was appointed professor of geometry at Gresham College in succession to Barrow. Dacres died in 1678.

Dacres was succeeded in his chair by Robert Hooke, and after the death of the latter in 1704 the chair was offered to **Andrew Tooke**. Tooke was born in London in 1673, took his B.A. degree from Clare in 1693, and died in 1731. He held the professorship until 1729, but with the beginning of the eighteenth century an appointment at Gresham College ceases to be a mark of scientific distinction.

The last of this trio was **Sir Samuel Morland**. Morland was born in Berkshire in 1625, and was educated at Winchester School and Magdalene College, but though he resided ten years at Cambridge he did not proceed to a degree. He took a prominent part in politics, and like most of his university contemporaries was a constitutional royalist. On the restoration he was made master of mechanics to the king, and thenceforward lived in or near London till his death on Jan. 6, 1696.

His earliest work on the quadrature of curves, partly printed in 1666, was at Pell's request withdrawn from publication—why, I do not know. In the same year he invented an admirable little arithmetical machine, an account of which was published in 1673. Morland seems subsequently to have turned his attention to the construction of machines. The speaking tube is one of his inventions: one of the first made was presented in 1671 to the library of Trinity College, and is still there. The form and construction of capstans, fire-engines, and certain other pumps were greatly improved by him, and the use of the barometer as a weather-gauge seems to be due to his advocacy. Some tables of interest, discount, and square and cube roots were also published by him at different dates after 1679.

CHAPTER IV.

THE LIFE AND WORKS OF NEWTON.

THE second occupant of the Lucasian chair was Newton. There is hardly a branch of modern mathematics, which cannot be traced back to him, and of which he did not revolutionize the treatment; and in the opinion of the greatest mathematicians of subsequent times—Lagrange, Laplace, and Gauss—his genius stands out without an equal in the whole history of mathematics. It will therefore be readily imagined how powerfully he must have impressed his methods and philosophy on the school which he suddenly raised to be the first in Europe; and the subsequent history of Cambridge (as far as this work is concerned therewith) is mainly that of the Newtonian philosophy.

Isaac Newton[1] was born in Lincolnshire near Grantham on Dec. 25, 1642 (O. S.), and died at Kensington, London, on March 20, 1727. He went to school at Grantham, and in 1661 came up as a subsizar to Trinity. Luckily he kept a diary, and we can thus form a fair idea of the reading of the best men at that time. He had not read any mathematics before coming into residence, but was acquainted with Sanderson's *Logic*, which was then frequently read as preliminary to

[1] The account in the text is condensed from chapter XVI. of my *History of mathematics*, London, 1888, to which I would refer the reader for authorities and fuller particulars. An edition of Newton's works was published by S. Horsley in 5 volumes, London, 1779—85: this contains a full bibliography of his writings.

mathematics. At the beginning of his first October term he happened to stroll down to Stourbridge Fair, and there picked up a book on astrology, but could not understand it on account of the geometry and trigonometry. He therefore bought a Euclid, and was surprised to find how obvious the propositions seemed. He thereupon read Oughtred's *Clavis* and Descartes's *Geometry*, the latter of which he managed to master by himself though with some difficulty. The interest he felt in the subject led him to take up mathematics rather than chemistry as a serious study. His subsequent mathematical reading as an undergraduate was founded on Kepler's *Optics*, the works of Vieta, Schooten's *Miscellanies*, Descartes's *Geometry*, and Wallis's *Arithmetica infinitorum*: he also attended Barrow's lectures. At a later time on reading Euclid more carefully he formed a very high opinion of it as an instrument of education, and he often expressed his regret that he had not applied himself to geometry before proceeding to algebraic analysis. He made some optical experiments and observations on lunar halos while an undergraduate. He was elected to a scholarship in 1663.

He took his B.A. degree in 1664. There is a manuscript of his written in the following year, and dated May 28, 1665, which is the earliest documentary proof of his discovery of fluxions. It was about the same time that he discovered the binomial theorem.

On account of the plague the college was sent down in the summer of 1665, and for the next year and a half Newton lived at home. This period was crowded with brilliant discoveries. He thought out the fundamental principles of his theory of gravitation, namely that every particle of matter attracts every other particle, and he suspected that the attraction varied as the product of their masses and inversely as the square of the distance between them. He also worked out the fluxional calculus tolerably completely: thus in a manuscript dated Nov. 13 of the same year he used fluxions to find the tangent and the radius of curvature at any point on a curve, and in October 1666 he applied them to several problems in the theory of equations. Newton communicated the results to his friends and pupils from and after 1669, but they were not published in print till many years later. It was also while staying at home at this time that he devised some instruments for grinding lenses to particular forms other than spherical, he perhaps decomposed light, and he certainly devoted considerable time to astrology and alchemy; indeed he never abandoned the idea of transmuting base metals into gold.

On his return to Cambridge in 1667 Newton was elected to a fellowship, and in 1668 took his M.A. degree. It is probable that he took pupils. His note-books shew that his attention was now mostly occupied with chemistry and optics, though there are a good many problems in pure and analytical geometry scattered amongst them.

During the next two years he revised and edited Barrow's *Lectures*, edited and added to Kinckhuysen's *Algebra*, and by using infinite series greatly extended the power of the method of quadratures given by Wallis. These however were only the fruits of his leisure; most of his time during these years being given up to optical researches.

In October 1669 Barrow had resigned the Lucasian chair in favour of Newton. Newton chose optics for the subject of his lectures and researches, and before the end of the year he had worked out the details of his discovery of the decomposition of a ray of white light into rays of different colours, which was effected by means of a prism bought at Stourbridge Fair. The complete explanation of the theory of the rainbow followed from this discovery. These discoveries formed the subject-matter of the lectures which he delivered as Lucasian professor in the years 1669, 1670, and 1671. The chief new results were embodied in papers published in the *Philosophical transactions* from 1671 to 1676. The manuscript of his original lectures was printed in 1729 under the title *Lectiones opticae*. This work is divided into two books, the first of which contains four sections and the second five. The first section of the first book deals with the decomposition of solar light by a prism in consequence of the unequal refrangibility of the rays that compose it, and gives a full account of his experiments. The second section contains an account of the method which Newton invented for determining the coefficients of refraction of different bodies. This is done by making a ray pass through a prism of the material so that the angle of incidence is equal to the angle of emergence: he shews that if the angle of the prism be i and the total deviation of the ray be δ the refractive index will be $\sin \frac{1}{2}(i+\delta) \operatorname{cosec} \frac{1}{2}i$. The third section is on refractions at plane surfaces. Most of this section is devoted to geometrical solutions of different problems, many of which are very difficult. He here finds the condition that a ray may pass through a prism with minimum deviation. The fourth section treats of refractions at curved surfaces. The second book treats of his theory of colours and of the rainbow.

By a curious chapter of accidents Newton failed to correct the chromatic aberration of two colours by means of a couple of prisms. He therefore abandoned the hope of making a refracting telescope which should be achromatic, and instead designed a reflecting telescope, probably on the model of a small one which he had made in 1668. The form he invented is that still known by his name. In 1672 he invented a reflecting microscope.

In 1675 he set himself to examine the problem as to how light was really produced. By the close of the year he had worked out the corpuscular or emission theory. Only three ways have been suggested in which light can be produced mechanically. Either the eye may be supposed to send out something which, so to speak, feels the object (as the Greeks believed); or the object perceived may send out something which hits or affects the eye (as Newton supposed); or there may be some medium between the eye and the object, and the object may cause some change in the form or nature of this intervening medium and thus affect the eye (as Huygens suggested in the wave or undulatory theory). It will be enough here to say that on either of the two latter theories all the obvious phenomena of geometrical optics such as reflexion, refraction, &c. can be accounted for. Within the present century crucial experiments have been devised which give different results according as one or the other theory is adopted; all these experiments agree with the results of the undulatory theory and differ from the results of the Newtonian theory: the latter is therefore untenable, but whether the former represents the whole truth and nothing but the truth is still an open question. Until however the theory of interference was worked out by Young the hypothesis of Huygens failed to account for all the facts and was open to more objections than that of Newton. Although Newton did not believe that the wave theory was the true explanation, he subsequently elaborated the fundamental principles of it.

His theory was embodied in two papers which were communicated to the Royal Society on Dec. 9 and Dec. 16 of 1672. In another paper on physical optics which was written in 1687 he elaborated the theory of fits of easy reflexion and transmission, the inflexion of light (bk. II. part 1), and the colours of thick plates (bk. II. part 4). The three papers together contain the whole of his emission theory of light, and comprise the great bulk of his treatise on optics published in 1704, to which the references given immediately above refer.

In 1673 he had written an account of his method of quadrature by means of infinite series in letters to Collins or Oldenburg; and in 1676 in answer to a request from Leibnitz he gave him a very brief account of his method and added the expansions of a binomial (i.e. the binomial theorem) and of $\sin^{-1} x$; from the latter of which he deduced that of $\sin x$. He also added an expression for the rectification of an elliptic arc in an infinite series.

Leibnitz wrote on Aug. 27, 1676, asking for fuller details, and on Oct. 24 Newton replied in a long but very interesting paper in which he gives an account of the way in which he had been led to some of his results.

He begins by saying that altogether he had used three methods for expansion in series. His first was arrived at from the study of the method of interpolation by which Wallis had found expressions for the area of the circle and hyperbola. Thus, by considering the series of expressions

$$(1-x^2)^{\frac{0}{2}},\ (1-x^2)^{\frac{2}{2}},\ (1-x^2)^{\frac{4}{2}},\ \&c.$$

he deduced by interpolations the law which connects the successive coefficients in the expansions of

$$(1-x^2)^{\frac{1}{2}},\ (1-x^2)^{\frac{3}{2}},\ \&c.$$

He then by analogy obtained the expression for the general term in the expansion of a binomial, i.e. the binomial theorem. He says that he proceeded to test this by forming the square of the expansion of $(1-x^2)^{\frac{1}{2}}$ which reduced to $1-x^2$; and he proceeded in a similar way with other expansions. He next tested the theorem in the case of $(1-x^2)^{\frac{1}{2}}$ by extracting the square root of $1-x^2$ *more arithmetico.* He also used the series to determine the areas of the circle and hyperbola in infinite series and found that they were the same as the results he had arrived at by other means.

Having established this result he then discarded the method of interpolation, and employed his binomial theorem as the most direct method of obtaining the areas and arcs of curves. Newton styled this his second method and it is the basis of his work on analysis by infinite series. He states that he had discovered it before the plague in 1665–66.

Newton then proceeds to state that he had also a third method; of which (he says) he had about 1669 sent an account to Barrow and Collins, illustrated by applications to areas, rectification, cubature, &c. This was the method of fluxions; but Newton gave no detailed description of it in this letter, probably because he thought that Leibnitz could, if he wished, obtain from Collins the explanation of it alluded to above. Newton added an anagram which described the method but which is unintelligible to any one to whom the key is not given. He gives however some illustrations of its use. The first is on the quadrature of the curves represented by

$$y = ax^m (b + cx^n)^p,$$

which he says can be determined as a sum of $(m+1)/n$ terms if $(m+1)/n$ be a positive integer, and which he thinks cannot otherwise be effected except by an infinite series. [This is not so, the integration is possible if $p + (m+1)/n$ be an integer.] He also gives a long list of other forms which are immediately integrable, of which the chief are

$$\frac{x^{mn-1}}{a + bx^n + cx^{2n}}, \quad \frac{x^{m(+\frac{1}{2})n-1}}{a + bx^n + cx^{2n}},$$

$$x^{mn-1} (a + bx^n + cx^{2n})^{\pm\frac{1}{2}},$$

$$x^{mn-1} (a + bx^n)^{\pm\frac{1}{2}} (c + dx^n)^{-1},$$

and $$x^{(m-1)n-1} (a + bx^n)^{\frac{1}{2}} (c + dx^n)^{-\frac{1}{2}};$$

where m is a positive integer and n is any number whatever.

At the end of his letter Newton alludes to the solution of the "inverse problem of tangents," a subject on which Leibnitz had asked for information. He gives formulæ for reversing any series, but says that besides these formulæ he has two methods for solving such questions which for the present he will not describe except by an anagram which being read is as follows, "Una methodus consistit in extractione fluentis quantitatis ex æquatione simul involvente fluxionem ejus. Altera tantum in assumptione seriei pro quantitate qualibet incognita ex qua cætera commode derivari possunt, et in collatione terminorum homologorum æquationis resultantis, ad eruendos terminos assumptæ seriei."

He adds in this letter that he is worried by the questions he is asked and the controversies raised about every new matter which he publishes, and he regrets that he has allowed his repose to be interrupted by running after shadows; and he implies that for the future he will publish nothing. As a matter of fact he did refuse to allow any account of his method of fluxions to be published till the year 1693.

Leibnitz did not reply to this letter till June 21, 1677. In his answer he explains his method of drawing tangents to curves, which he says proceeds "not by fluxions of lines but by the differences of numbers"; and he introduces his notation of dx and dy for the infinitesimal differences between the co-ordinates of two consecutive points on a curve. He also gives a solution of the problem to find a curve whose subtangent is constant, which shews that he could integrate.

I do not know with any certainty on what subjects Newton was chiefly occupied during the next eight years, 1676—1684. He was partly engaged in chemical experiments and partly in geological speculations; and I believe that his experiments in electricity and magnetism and the law of cooling in the theory of heat are of this date. A large part of the geometry and the pure mathematics subsequently incorporated in the first book of the *Principia* should probably be also referred to this time; and perhaps some parts of the essay on cubic curves.

It is almost certain that the *Universal arithmetic* which is on algebra, theory of equations, and miscellaneous problems contains the substance of Newton's lectures during these years. His manuscript of it is still extant. Amongst several new theorems on various points in algebra and the theory of equations the following important results were here first enunciated. He explained that the equation whose roots are the solution of a given problem will have as many roots as there are different possible cases; and he also considered how it happened that the equation to which a problem led might contain roots which did not satisfy the original question. He extended Descartes's rule of signs to give limits to the number of imaginary roots. He used the principle of continuity to explain how two real and unequal roots might become imaginary in passing through equality, and illustrated this by geometrical considerations; thence he shewed that imaginary roots must occur in pairs. Newton also here gave rules to find a superior limit to the positive roots of a numerical equation, and to determine the approximate values of the numerical roots. He further enunciated the theorem known by his name for finding the sum of the nth powers of the roots of an equation, and laid the foundation of the theory of symmetrical functions of the roots of an equation.

In August 1684 Newton received a visit from Halley who drew his attention to the motion of the moon. Hooke, Huygens, Halley, and Wren had all conjectured that the force of the attraction of the sun or earth on an external particle varied inversely as the square of the distance. These writers seem to have independently shewn that if Kepler's conclusions were rigorously true, as to which they were not quite certain, the law of attraction must be that of the inverse square, but they could not deduce from the law the orbits of the planets. When Halley visited Cambridge in August 1684 he explained that their investigations were stopped by their inability to solve this problem, and asked Newton if he could find out what the orbit of a planet would be if the law of attraction were that of the inverse square. Newton immediately replied that it was an ellipse, and promised to send or write out afresh a demonstration of it which he had given in 1679. This was sent in November 1684.

Instigated by this question, Newton now attacked the whole problem of gravitation, and succeeded in shewing that if the distances of the members of the solar system were so great that they might for the purpose of their mutual attraction be regarded as points then their motions were in accordance with the law of gravitation. The elements of these discoveries were put together in the tract called *De motu*, which contains the substance of sections ii. and iii. of the first book of the *Principia*, and was read by Newton for his lectures in the Michaelmas term 1684.

Newton however had not yet determined the attraction of

a spherical body on any external point, nor had he calculated the details of the planetary motions even if the members of the solar system could be regarded as points. The first problem was solved at the latest in February 1685. Till he had effected this his theory had been shewn to be true only in so far as the sun can be regarded as a point compared with its distance from the planets, or the earth as a point compared with its distance from the moon; but this discovery shewed that it was mathematically true, excepting only for the slight deviation from a perfectly spherical form of the sun, earth and planets. It was thus now in his power to apply mathematical analysis with absolute precision to the explanation of the detailed phenomena of the solar system. This he did in the almost incredibly short space of time from March 1686 to the end of March 1687, and the result is embodied in the *Principia*[1]. Of the three fundamental principles there applied we may say that the idea that every particle attracts every other particle in the universe was formed at least as early as 1666; the law of equable description of areas, its consequences, and the fact that if the law of attraction were that of the inverse square the orbit of a particle about a centre of force would be a conic were proved in 1679; and lastly the discovery that a sphere, whose density at any point depends only on the distance from the centre, attracts an external point as if the whole mass were collected at its centre was made in 1685. It was this last discovery that enabled him to apply the first two principles to the phenomena of bodies of finite size.

The first book of the *Principia* was finished on April 28, 1686. This book is given up to the consideration of the motion of particles or bodies in free space either in known orbits, or under the action of known forces, or under their mutual attraction. In it Newton generalizes the law of attraction into a statement that every particle of matter in the

[1] A brief analysis of the subject-matter of the *Principia* is given on pp. 310—21 of my *History of mathematics*, London, 1888.

universe attracts every other particle with a force which varies directly as the product of their masses and inversely as the square of the distance between them; and he thence deduces the law of attraction for spherical shells of constant density. The book is prefaced by an introduction on the science of dynamics.

In another three months, that is by the summer of 1686, he had finished the second book of the *Principia*. This book treats of motion in a resisting medium, and of hydrostatics and hydrodynamics, with special applications to waves, tides, and acoustics. He concludes it by shewing that the Cartesian theory of vortices was inconsistent both with the known facts and with the laws of motion.

The next nine or ten months were devoted to the third book. For this he probably had no materials ready. In it the theorems obtained in the first book are applied to the chief phenomena of the solar system, the masses and distances of the planets and (whenever sufficient data existed) of their satellites are determined. In particular the motion of the moon, the various inequalities therein, and the theory of the tides are worked out in great detail. He also investigates the theory of comets, shews that they belong to the solar system, explains how from three observations the orbit can be determined, and illustrates his results by considering certain special comets. The third book as we have it is but little more than a sketch of what Newton had proposed to himself to accomplish. The original programme of the work is extant and his note-books shew that he continued to work at it for some years after the publication of the first edition of the *Principia*.

The printing of the work was very slow and it was not finally published till the summer of 1687. The conciseness, absence of illustrations, and synthetical character of the book as first issued seriously restricted the numbers of those who were able to appreciate its value; and though nearly all competent critics admitted the validity of the conclusions a considerable time elapsed before it affected the current beliefs of educated men. I should be inclined to say (but on this point opinions differ widely) that within ten years of its publication it was generally accepted in Britain as giving a correct account of the laws of the universe; it was similarly accepted within about twenty years on the continent, except in France where patriotism was urged in defence of the Cartesian theory until Voltaire in 1738 took up the advocacy of the Newtonian theory.

The manuscript of the *Principia* was finished by 1686. Newton devoted the remainder of that year to his paper on physical optics, the greater part of which is given up to the subject of diffraction (see p. 55).

In 1687 James II. having tried to force the university to admit as a master of arts a Roman Catholic priest who refused to take the oaths of supremacy and allegiance, Newton took a prominent part in resisting the illegal interference of the king, and was one of the deputation sent to London to protect the rights of the university. The active part taken by Newton in this affair led to his being in 1689 elected member for the university. This parliament only lasted thirteen months, and on its dissolution he gave up his seat. At a later date he was returned on one or two occasions, but he never took any prominent part in politics.

On his coming back to Cambridge in 1690 he resumed his mathematical studies and correspondence. If he lectured at this time (which is doubtful) it was on the subject-matter of the *Principia*. The two letters to Wallis in which he explained his method of fluxions and fluents were written in 1692, and were published in 1693. Towards the close of 1692 and throughout the two following years Newton had a long illness, suffering from insomnia and general nervous irritability. He never quite regained his elasticity of mind, and though after his recovery he shewed the same power in solving any question propounded to him, he ceased thenceforward to do original work on his own initiative, and it was difficult to stir him to activity.

In 1694 Newton began to collect data connected with the irregularities of the moon's motion with the view of revising the part of the *Principia* which dealt with that subject. To render the observations more accurate he forwarded to Flamsteed a table which he had previously made of corrections for refraction. This was not published till 1721 when Halley communicated it to the Royal Society. The original calculations of Newton and the papers connected with it are in the Portsmouth collection at Cambridge, and shew that Newton obtained it by finding the path of a ray by means of quadratures in a manner equivalent to the solution of a differential equation. As an illustration of Newton's genius I may mention that even as late as 1754 Euler failed to solve the same problem. In 1782 Laplace gave a rule for constructing the table, and his results agree substantially with those of Newton.

I do not suppose that Newton would in any case have produced much more original work after his illness; but his appointment in 1695 as warden, and his promotion in 1699 to the mastership of the mint at a salary of £1500 a year, brought his scientific investigations to an end. He now moved to London. In 1701 he resigned the Lucasian chair, and in 1703 he was elected president of the Royal Society.

In 1704 he published his *Optics*, containing an account of his emission theory of light (see p. 55). To this book two appendices were added; one on cubic curves, and the other on the quadrature of curves and his theory of fluxions. Both of these were old manuscripts which had long been known to his friends at Cambridge, but had been previously unpublished.

The first of these appendices is entitled *Enumeratio linearum tertii ordinis* and was apparently written before 1676. The object seems to be to illustrate the use of analytical geometry, and as the application to conics was well known Newton selected the theory of cubics.

He begins with some general theorems, and classifies curves according as to whether their equations are algebraical or transcendental: the former being cut by a straight line in a number of points (real or imaginary) equal to the degree of the curve, the latter being cut by a straight line in an infinite number of points. Newton then shews that many of the most important properties of conics have their analogues in the theory of cubics; of this he gives numerous illustrations. He next proceeds to discuss the theory of asymptotes and curvilinear diameters to curves of any degree.

After these general theorems he commences his detailed examination of cubics by pointing out that a cubic must at least have one real asymptotic direction. If the asymptote corresponding to this direction be at a finite distance it may be taken for the axis of y. This asymptote will cut the curve in three points altogether, of which at least two are at infinity. If the third point be at a finite distance then (by one of his general theorems on asymptotes) the equation can be written in the form

$$xy^2 + hy = ax^3 + bx^2 + cx + d,$$

while if the third point in which this asymptote cuts the curve be also at infinity the equation can be written in the form

$$xy = ax^3 + bx^2 + cx + d.$$

Next he takes the case where the asymptote corresponding to the real asymptotic direction is not at a finite distance. A line parallel to it may be taken as the axis of y. Any such line will cut the curve in three points altogether, of which one is by hypothesis at infinity, and one is necessarily at a finite distance. He then shews that if the remaining point in which this line cuts the curve be at a finite distance the equation can be written in the form

$$y^2 = ax^3 + bx^2 + cx + d,$$

while if it be at an infinite distance the equation can be written in the form

$$y = ax^3 + bx^2 + cx + d.$$

Any cubic is therefore reducible to one of four characteristic forms. Each of these forms is then discussed in detail, and the possibility of the existence of double points, isolated ovals, &c. is thoroughly worked out. The final result is that there are in all seventy-two possible forms which a cubic may take. To these Stirling in his *Lineae tertii ordinis Newtonianae* published in 1717 added four; and Cramer and Murdoch in the *Genesis curvarum per umbras* published in 1746 each added one; thus making in all seventy-eight species. In the course of the analysis Newton states the remarkable theorem that in the same way as the conics may be considered as the shadows of a circle (i.e. plane sections of a cone on a circular base) so all cubics may be considered as the shadows of the curves represented by the equation $y^2 = ax^3 + bx^2 + cx + d$.

The second appendix to the *Optics* was entitled *De quadratura curvarum.* Most of it had been communicated to Barrow in 1666, and was probably familiar to Newton's pupils and friends from about 1667 onwards. It consists of two parts.

The bulk of the first part had been included in the letter to Leibnitz of Oct. 24, 1676. This part contains the earliest use of literal indices, and the first printed statement of the binomial theorem: these are however introduced incidentally. The main object of this part is to give rules for developing a function of x in a series in ascending powers of x; so as to enable mathematicians to effect the quadrature of any curve in which the ordinate y can be expressed as an explicit function of the abscissa x. Wallis had shewn how this quadrature could be found when y was given as a sum of a number of powers of x (see p. 43), and Newton here extends this by shewing how any function can be expressed as an infinite series in that way. I should add that Newton is generally careful to state whether the series are convergent. In this way he effects the quadrature of the curves

$$y = \frac{a^2}{b+x}, \quad y = (a^2 \pm x^2)^{\frac{1}{2}}, \quad y = (x - x^2)^{\frac{1}{2}}, \quad y = \left(\frac{1+ax^2}{1-bx^2}\right)^{\frac{1}{2}},$$

but the results are of course expressed as infinite series. He then proceeds to curves whose ordinate is given as an implicit function of the abscissa: and he gives a method by which y can be expressed as an infinite series in ascending powers of x, but the application of the rule to any curve demands in general such complicated numerical calculations as to render it of little value. He concludes this part by shewing that the rectification of a curve can be effected in a somewhat similar way. His process is equivalent to finding the integral with regard to x of $(1+\dot{y}^2)^{\frac{1}{2}}$ in the form of an infinite series.

This part should be read in connection with his *Analysis by infinite series* published in 1711, and his *Methodus differentialis* published in 1736. Some additional theorems are there given, and in the latter of these works he discusses his method of interpolation. The principle is this. If $y=\phi(x)$ is a function of x and if when x is successively put equal to a_1, a_2, ... the values of y are known and are b_1, b_2 ... then a parabola whose equation is $y=p+qx+rx^2+\ldots$ can be drawn through the points (a_1, b_1), (a_2, b_2), ... and the ordinate of this parabola may be taken as an approximation to the ordinate of the curve. The degree of the parabola will of course be one less than the number of given points. Newton points out that in this way the areas of any curves can be approximately determined.

The second part of this second appendix contains a description of his method of fluxions and is condensed from his manuscript to which allusion is made a few pages later (see p. 70).

The remaining events of Newton's life may be summed up very briefly. In 1705 he was knighted. From this time onwards he devoted much of his leisure to theology, and wrote at great length on prophecies and predictions which had always been subjects of interest to him. His *Universal arithmetic* was published by Whiston in 1707, and his *Analysis by infinite series* in 1711; but Newton had nothing to do with preparing either of these for the press. In 1709 Newton was persuaded to allow Cotes to prepare the long-talked-of second edition of the *Principia*; it was issued in March 1713. A third edition was published in 1726 under the direction of Henry Pemberton. Newton's original manuscript on fluxions was published in 1736, some nine years after his death, by John Colson. In 1725 his health began to fail. He died on March 20, 1727, and eight days later was buried with great state in Westminster Abbey.

In appearance Newton was short, and towards the close of his life rather stout, but well set, with a square lower jaw, a very broad forehead, rather sharp features, and brown eyes. His hair turned grey before he was thirty, and remained thick and white as silver till his death. He dressed in a slovenly manner, was rather languid, and was generally so absorbed in his own thoughts as to be anything but a lively companion.

Many anecdotes of his extreme absence of mind when engaged in any investigation have been preserved. Thus once when riding home from Grantham he dismounted to lead his horse up a steep hill, when he turned at the top to remount he found that he had the bridle in his hand, while his horse had slipped it and gone away. Again on the few occasions when he sacrificed his time to entertain his friends, if he left them to get more wine or for any similar reason, he would as often as not be found after the lapse of some time working out a problem, oblivious alike of his expectant guests and of his errand. He took no exercise, indulged in no amusements, and worked incessantly, often spending 18 or 19 hours out of the 24 in writing. He modestly attributed his discoveries largely to the admirable work done by his predecessors; and in answer to a correspondent he explained that if he had seen farther than other men, it was only because he had stood on the shoulders of giants. He was morbidly sensitive to being involved in any discussions. I believe that with the exception of his two papers on optics in 1675, every one of his works was only published under pressure from his friends and against his own wishes. There are several instances of his communicating papers and results on condition that his name should not be published.

In character he was perfectly straightforward and honest, but in his controversies with Leibnitz, Hooke, and others though scrupulously just he was not generous. During the early half of his life he was parsimonious, if not stingy, and he was never liberal in money matters.

The above account, slight though it is, will yet enable the reader to form an idea of the immense extent of Newton's services to science. His achievements are the more wonderful if we consider that most of them were effected within twenty-five years, 1666—1692. Two branches of applied mathematics stand out pre-eminent in his work: first, his theories of physical and geometrical optics; and second, his theory of gravitation or physical astronomy. Although unrivalled in his power of analysis—of which his *Universal arithmetic* and the essay on cubic curves would alone be sufficient evidence—he always by choice presented his proofs in a geometrical form. But it is known that for purposes of research he generally used the fluxional calculus in the first instance. Hence excessive importance was attached by the Newtonian school to these two branches of pure mathematics. So completely did Newton impress his individuality on English mathematics that during the eighteenth century the subject at Cambridge meant little else but a study of the four branches above mentioned. I have already alluded to the subject-matter of the *Principia* and *Optics,* and I must now say a few words on his method of exposition, and his use of geometry and fluxions.

It is probable that no mathematician has ever equalled Newton in his command of the processes of classical geometry. But his adoption of it for purposes of demonstration appears to have arisen from the fact that the infinitesimal calculus was then unknown to most of his readers, and had he used it to demonstrate results which were in themselves opposed to the prevalent philosophy of the time the controversy would have first turned on the validity of the methods employed. Newton therefore cast the demonstrations of the *Principia* into a geometrical shape which, if somewhat longer, could at any rate be made intelligible to all mathematical students and of which the methods were above suspicion. In further explanation of this I ought to add that in Newton's time and for nearly a century afterwards the differential and fluxional calculus were not fully developed and did not possess the same superiority over the method he adopted which they do now. The effect of his confining himself rigorously to classical geometry and elementary algebra, and of his refusal to make any use even of analytical geometry and of trigonometry is that the *Principia* is written in a language which is archaic (even if not unfamiliar) to us. The subject of optics lends itself more readily to a geometrical treatment, and thus his demonstrations of theorems in that subject are not very different to those still used.

The adoption of geometrical methods in the *Principia* for purposes of demonstration does not indicate a preference on Newton's part for geometry over analysis as an instrument of research, for it is now known that Newton used the fluxional calculus in the first instance in finding some of the theorems (especially those towards the end of book I. and in book II.), and then gave geometrical proofs of his results. This translation of numerous theorems of great complexity into the language of the geometry of Archimedes and Apollonius is I suppose one of the most wonderful intellectual feats which was ever performed.

The fluxional calculus is one form of the infinitesimal calculus expressed in a certain notation just as the differential calculus is another aspect of the same calculus expressed in a different notation. Newton assumed that all geometrical magnitudes might be conceived as generated by continuous motion: thus a line may be considered as generated by the motion of a point, a surface by that of a line, a solid by that of a surface, a plane angle by the rotation of a line, and so on. The quantity thus generated was defined by him as the fluent or flowing quantity. The velocity of the moving magnitude was defined as the fluxion of the fluent.

The following is a summary of Newton's treatment of fluxions. There are two kinds of problems. The object of the first is to find the fluxion of a given quantity, or more generally "the relation of the fluents being given to find the relation of their fluxions." This is equivalent to differentiation. The object of the second or inverse method of fluxions is from the fluxion or some relation involving it to determine the fluent, or more generally "an equation being proposed exhibiting the relation of the fluxions of quantities to find the relations of those quantities or fluents to one another[1]." This is equivalent either to integration which Newton termed the method of quadrature, or to the solution of a differential equation which was called by Newton the inverse method of tangents. The methods for solving these problems are discussed at considerable length.

Newton then went on to apply these results to questions connected with the maxima and minima of quantities, the method of drawing tangents to curves, and the curvature of curves (viz. the determination of the centre of curvature, the radius of curvature, and the rate at which the radius of curvature increases). He next considered the quadrature of curves and the rectification of curves[2].

It has been remarked that neither Newton nor Leibnitz produced a calculus, that is a classified collection of rules; and that the problems they discussed were treated from first principles. That no doubt is the usual sequence in the history of such discoveries, though the fact is frequently forgotten by subsequent writers. In this case I think the statement, so far as Newton is concerned, is incorrect, as the foregoing account sufficiently shews.

If a flowing quantity or fluent were represented by x, Newton

[1] Colson's edition of Newton's manuscript, pp. xxi. xxii.

[2] Colson's edition of Newton's manuscript, pp. xxii. xxiii.

denoted its fluxion by $\dot{x}$, the fluxion of $\dot{x}$ or second fluxion of x by $\ddot{x}$, and so on. Similarly the fluent of x was denoted by x' or $[x]$ or $\boxed{x}$. The infinitely small part by which a fluent such as x increased in a small interval of time measured by o was called the moment of the fluent; and its value was shewn to be $\dot{x}o$[1]. I should here note the fact that Vince and other writers in the eighteenth century used $\dot{x}$ to denote the increment of x and not the velocity with which it increased; that is $\dot{x}$ in their writings stands for what Newton would have expressed by $\dot{x}o$ and what Leibnitz would have written as dx. They also used the current symbol for integration. Thus $\int x^n \dot{x}$ stands with them for what Newton would have usually expressed by $\boxed{x^n}$, or what Leibnitz would have written as $\int x^n dx$.

I need not here concern myself with the details as to how Newton treated the problems above mentioned. I will only add that in spite of the form of his definition the introduction in geometry of the idea of time was evaded by supposing that some quantity (e.g. the abscissa of a point on a curve) increased equably; and the required results then depend on the rate at which other quantities (e.g. the ordinate or radius of curvature) increase relatively to the one so chosen[2]. The fluent so chosen is what we now call the independent variable; its fluxion was termed the "principal fluxion;" and of course if it were denoted by x then $\dot{x}$ was constant, and consequently $\ddot{x} = 0$.

Newton's manuscript, from which most of the above summary has been taken, is believed to have been written between 1671 and 1677, and to have been in circulation at Cambridge from that time onwards. It was unfortunate that it was not published at once. Strangers at a distance naturally judged of the method by the letter to Wallis in 1692 or the *Tractatus de*

[1] Colson's edition of Newton's manuscript, p. 24.

[2] Colson's edition of Newton's manuscript, p. 20.

quadratura curvarum, and were not aware that it had been so completely developed at an earlier date. This was the cause of numerous misunderstandings.

The notation of the fluxional calculus is for most purposes less convenient than that of the differential calculus. The latter was invented by Leibnitz in 1675, and published in 1684. But the question whether the general idea of the calculus expressed in that notation was obtained by Leibnitz from Newton or whether it was invented independently gave rise to a long and bitter controversy. From what I have read of the voluminous literature on the question, I think on the whole it points to the fact that Leibnitz obtained the idea of the differential calculus from a manuscript of Newton's which he saw in 1673, but the question is one of considerable difficulty and no one now is likely to dogmatize on it[1].

If we must confine ourselves to one system of notation then there can be no doubt that that which was designed by Leibnitz is better fitted for most of the purposes to which the infinitesimal calculus is applied than that of fluxions, and for some (such as the calculus of variations) it is indeed almost essential. His form of the infinitesimal calculus was adopted by all continental mathematicians. In England the controversy with Leibnitz was regarded as an attempt by foreigners to defraud Newton of the credit of his invention, and the question was complicated on both sides by national jealousies. It was therefore natural though it was unfortunate that the geometrical and fluxional methods (as used by Newton) should be alone studied and employed at Cambridge. For more than a century the English school was thus quite out of touch with continental mathematicians. The consequence was that

[1] The case in favour of the independent invention by Leibnitz is stated in Biot and Lefort's edition of the *Commercium epistolicum*, Paris, 1856, and in an article in the *Philosophical magazine* for 1852. A summary of the arguments on the other side is given in Dr Sloman's *The claims of Leibnitz to the invention of the differential calculus* issued at Leipzig in 1858, of which an English translation was published at Cambridge in 1860.

in spite of the brilliant band of scholars formed by Newton the improvements in the methods of analysis gradually effected on the continent were almost unknown in Cambridge. It was not until about 1820 (as described in chapter VII.) that the value of analytical methods was fully recognized in England; and that Newton's countrymen again took any large share in the developement of mathematics.

CHAPTER V.

THE RISE OF THE NEWTONIAN SCHOOL.

CIRC. 1690—1730.

In the last chapter I enumerated very briefly the more important discoveries of Newton, and pointed out the four subjects to which he paid special attention. I have now to describe how those discoveries affected the study of mathematics in the university, and led to the rise of the Newtonian school.

The mathematical school in the university prior to Newton's time contained several distinguished men, but in point of numbers it was not large. We need not therefore be surprised to find that it was Newton's theory of the universe and not his mathematics that excited most attention in the university; and it was because mathematics supplied the key to that theory that it began to be studied so eagerly. Hence the rise of the Newtonian school dates from the publication of the *Principia*.

In considering the history of this school, it must be remembered that at Cambridge until recently professors only rarely put themselves into contact with or adapted their lectures for the bulk of the students in their own department. Accordingly if we desire to find to whom the spread of a general study of the Newtonian philosophy was immediately due, we must look not to Newton's lectures or writings, but among those proctors, moderators, or college tutors, who had accepted his doctrines. The form in which the *Principia* was cast, its extreme conciseness, the absence of all illustrations, and the immense interval between the abilities of Newton and those of his contemporaries combined to delay the acceptance of the new philosophy; and it is a matter of surprise that its truth was so soon recognized.

I propose first to mention Richard Laughton, Samuel Clarke, John Craig, and John Flamsteed, who were some of the earliest residents to accept the Newtonian philosophy. I must then devote a few words to Bentley, to whom the predominance in the university of the Newtonian school is largely due: he knew but little mathematics himself, but he used his considerable influence to put the study on a satisfactory basis. I shall then briefly describe the works of William Whiston, Nicholas Saunderson, Thomas Byrdall, James Jurin, Brook Taylor, Roger Cotes, and Robert Smith: the three mathematicians last named being among the most powerful of Newton's immediate successors. Lastly I propose to describe the course of reading in mathematics of a student at Cambridge about the year 1730, which I take as the limit of the period treated in this chapter.

Among the earliest of those who realized the importance of Newton's discoveries was **Richard Laughton**[1], a fellow of Clare Hall. I have been unable to discover any account of his life, but I find he is referred to as the most celebrated "pupil-monger" of his time, and I gather from references to him in the literature of the period that he was one of the most influential of those who introduced a study of the Newtonian theory of the universe into the university curriculum. In 1694 he persuaded Samuel Clarke (who was probably one of his pupils) to defend in the schools a question on physical astronomy taken from the *Principia*, and in the same year the Cartesian theory was ridiculed in the tripos verses. These seem to be the earliest allusions in the public exercises

[1] The name was pronounced Laffton: see Uffenbach's account of his visit to Cambridge in 1710 quoted on p. 6 of the *Scholae academicae*.

of the university to the Newtonian philosophy; but so rapidly were its merits appreciated that within twenty years it was the dominant study in the university. Later in the same year Laughton was made a tutor of Clare; and thenceforward he took every opportunity of his new position to urge his pupils to read Newton.

In 1710 Laughton was proctor, and claimed the right to preside in person at the acts in the schools. This was a part of the ancient duties of the office, but since 1680 it had been customary for the senate each year to appoint moderators who performed it as the deputies of the proctors, and even at an earlier date it was not unusual for the latter officers to select moderators (or posers, as they were then generally designated) to whom they delegated that part of their work. Laughton presided in person, and in summing up the discussions exposed the assumptions and mistakes in the Cartesian system. A resident[1] who was no special advocate of the new doctrines bears witness in his diary to the success of Laughton's efforts. "It is certain," says he, "that for some years [before 1710] he had been diligently inculcating [the Newtonian] doctrines, and that the credit and popularity of his college had risen very high in consequence of his reputation." Acting as proctor in that year Laughton induced William Browne of Peterhouse to keep his acts on mathematical questions, and promised him an honorary proctor's optime degree (see p. 170) if he would do so. Laughton died in 1726.

The earliest text-book with which I am acquainted written to advocate the Newtonian philosophy was by the Samuel Clarke to whom allusion has just been made. **Samuel Clarke**[2] was born at Norwich on Oct. 11, 1675, and took his B.A. from Caius in 1695. The text-book on physical astronomy then in common use was Rohault's *Physics*, which was

[1] See the *Diary of Ralph Thorseby* (1677—1724) edited by J. Hunter, 2 volumes, London, 1830.

[2] See his life and works by B. Hoadly, 4 volumes, London, 1738; and a memoir by W. Whiston, third edition, London, 1741.

founded on Descartes's hypothesis of vortices. Clarke thought that he could best advocate the Newtonian theory by issuing a new edition of Rohault with notes, shewing that the conclusions were necessarily wrong. This curious mixture of truth and falsehood continued to be read at Cambridge at least as late as 1730, and went through several editions. After 1697 Clarke devoted most of his time to the study of theology, though in 1706 he translated Newton's *Optics* into "elegant Latin," with which Newton was so pleased that he sent him a present of five hundred guineas. In 1728 Clarke contributed a paper to the *Philosophical transactions* on the controversy then raging as to whether a force ought to be measured by the momentum or by the kinetic energy produced in a given mass. He died in 1729.

Another mathematician of this time who did a good deal to bring fluxions into general use was Craig. **John Craig** was born in Scotland. He came to Cambridge about 1680, but it is believed he never took a degree. He went down in 1708, and after holding various livings settled in London, where he died on Oct. 11, 1731. His chief works were the *Methodus...quadraturas determinandi* published in 1685, the *De figurarum quadraturis et locis geometricis* published in 1693, and the *De calculo fluentium* (2 volumes) and *De optica analytica* (2 volumes) which were published in 1718. In the two works first mentioned he argues in favour of the ideas and notation of the differential calculus, and in connection with them he had a long controversy with Jacob Bernoulli. In the last he definitely adopts the fluxional calculus as the correct way of presenting the truths of the infinitesimal calculus. These works shew that Craig was a good mathematician.

Among his papers published in the *Philosophical transactions* I note one in 1698 on the quadrature of the logarithmic curve, one in 1700 on the curve of quickest descent, and another in the same year on the solid of least resistance, one in 1703 on the quadrature of any curve, one in 1704 containing a solution of a problem issued by John Bernoulli as a challenge, one in 1708 on the rectification of any curve, and lastly one in 1710 on the construction of logarithmic tables.

It is however much easier to obtain a lasting reputation by eccentricity than by merit; and hundreds who never heard of Craig's work on fluxions know of him as the author of *Theologia Christianae principia mathematica* published in 1699. He here starts with the hypothesis[1] that evidence transmitted through successive generations diminishes in credibility as the square of the time. The general idea was due to the Mahommedan apologists, who enunciated it as an axiom, and then argued that as the evidence for the Christian miracles daily grows weaker a time must come when they will have no evidential value, whence the necessity of another prophet. Curiously enough Craig's formulæ shew that the oral evidence would by itself have become worthless in the eighth century, which is not so very far removed from the date of Mahommed's death (632). He asserts that the gospel evidence will cease to have any value in the year 3150. He then quotes a text to shew that at the second coming faith will not be quite extinct among men: and hence the world must come to an end before 3150. This was reprinted abroad, and seriously answered by many divines; but most of his opponents were better theologians than mathematicians, and would have been wiser if they had contented themselves with denying his axioms.

I must not pass over this period without mentioning Flamsteed. **John Flamsteed**[2] was born in Derbyshire in 1646. When at school he picked up a copy of Holywood's treatise on the sphere (see p. 5) and was so fascinated by it that he determined to study astronomy. It was intended to send him to Cambridge, but for some years he was too delicate to leave home. He however obtained copies of Street's *Astronomy*, Riccioli's *Almagestum novum*, and Kepler's *Tables*, which he read by himself. By the time he was twenty-two or three he

[1] See pp. 77, 78 of *A budget of paradoxes* by A. De Morgan, London, 1872.

[2] See his life, by R. F. Baily, London, 1835.

was already one of the best astronomers (both theoretical and practical) in Europe. He entered at Jesus College in 1670, and devoted himself to the study of mathematics, optics, and astronomy. He seems to have been in constant communication with Barrow and Newton. He took his B.A. in 1674, and in the following year was appointed to take charge of the national observatory then being erected at Greenwich. He is thus the earliest of the astronomer-royals. He gave Newton many of the data for the numerical calculations in the third book of the *Principia*, but in consequence of a quarrel, refused to give the additional ones required for the second edition. He died at Greenwich in 1719.

He invented the system (published in 1680) of drawing maps by projecting the surface of the sphere from the centre on an enveloping cone which can then be unwrapped. He wrote papers on various astronomical problems, but his great work, which is an enduring memorial of his skill and genius, is his *Historia coelestis Brittanica*, edited by Halley and published posthumously in three volumes in 1725.

By the beginning of the eighteenth century the immense reputation and great powers of Newton were everywhere recognized. The adoption of his methods and philosophy at Cambridge was however in no slight degree due to other than professed mathematicians. Of these the most eminent was Bentley, who invariably exerted his influence to make literature and mathematical science the distinctive features of a Cambridge training. Philosophy was also still read and was not unworthily represented by Bacon, Descartes, and Locke[1]. It was from

[1] *Francis Bacon*, born in 1561, was educated at Trinity College, Cambridge, and died in 1626: the *Novum organum* was published in 1620. *René Descartes* was born in 1596 and died in 1650: his *Discours* was published in 1637, and his *Meditations* in 1641. *John Locke*, born in 1632, was educated at Christ Church, Oxford, and died in 1704: his *Essay concerning human understanding* was published in 1690.

Newton aided by Bentley that the Cambridge of the eighteenth century drew its inspiration, and it was their influence that made the intellectual life of the university during that time so much more active than that of Oxford.

Richard Bentley[1] was born in Yorkshire on Jan. 27, 1662, and died at Cambridge on July 14, 1742. He took his B.A. from St John's College in 1680 as third wrangler, but in consequence of the power of conferring honorary optime degrees (see p. 170) his name appears as sixth in the list. He was not eligible for a fellowship, and in 1682 went down.

In 1692 he was selected to deliver the first course of the Boyle lectures on theology, which had been founded by the will of Robert Boyle, who died in 1691. In the sixth, seventh, and eighth sermons he gave a sketch of the Newtonian discoveries: this was expressed in non-technical language and excited considerable interest among those members of the general public who had been unable to follow the mathematical form in which Newton's arguments and investigations had been previously expressed.

In 1699 Bentley was appointed master of Trinity College, and from that time to his death an account of his life is the history of Cambridge. It is almost impossible to overrate his services to literature and scientific criticism, and his influence on the intellectual life of the university was of the best. It is however indisputable that many of his acts were illegal, and the fact that he wished to promote the interests of learning is no excuse for the arrogance, injustice, and tyranny which characterized his rule.

One reform of undoubted wisdom which he introduced may

[1] See the *Life of Bentley* by W. H. Monk, 2 vols., London, 1833: see also the volume by R. C. Jebb in the series of *English men of letters*, London, 1882; the latter on the whole is eulogistic, and it must be remembered that most of Bentley's Cambridge contemporaries would not have taken so favourable a view of his character. Another brilliant monograph on Bentley from the pen of Hartley Coleridge will be found in the *Worthies of Yorkshire and Lancashire*, London, 1836.

be here mentioned. Elections to scholarships and fellowships at that time took place on the result of a *viva voce* examination by the master and seniors in the chapel. To give an opportunity for written exercises and time for discussion by the electors of the merits of the candidates, Bentley arranged that every candidate should be first examined by each elector. In practice part of the examination was always oral and part written. He also made the award of scholarships annual instead of biennial, and admitted freshmen to compete for them. In 1789 the examination was made the same for all candidates and conducted openly. A survival of the old practice—after nearly two hundred years—exists in the fact that the electors to fellowships and scholarships still always adjourn to the chapel to make the technical election and declaration.

The following account of the scholarship examination for 1709 taken from a letter[1] of one of the candidates (John Byrom) to his father may interest the reader, as it is the earliest account of such an examination which I have seen. In that year there were apparently ten vacancies, and nineteen students "sat" for them. At the end of April every candidate sent a letter in Latin to the master and each of the seniors announcing that he should present himself for the examination. On May 7 Byrom was examined by the vice-master, on the following Monday and Tuesday he was examined by Bentley, Stubbs, and Smith in their respective rooms, and on Wednesday he went to the lodge and while there wrote an essay: the other seniors seem to have shirked taking part in the examination. "On Thursday," writes Byrom, "the master and seniors met in the chapel for the election; Dr Smith had the gout and was not there. They stayed consulting about an hour and a half, and then the master wrote the names of the elect, who (*sic*) shewed me mine in the list. Fifteen were chosen. [The

1 See p. 6 of the *Remains of John Byrom*, Chetham Society Publications, Manchester, 1854.

five lowest being pre-elected to the next vacancies]....Friday noon we went to the master's lodge, where we were sworn in in great solemnity, the senior Westminster reading the oath in Latin, all of us kissing the Greek Testament. Then we kneeled down before the master, who took our hands in his and admitted us scholars in the name of the Father, Son, &c. Then we went and wrote our names in the book and came away, and to-day gave in our epistle of thanks to the master. We took our places at the scholars' table last night. To-day the new scholars began to read the lessons in chapel and wait [i.e. to read grace] in the hall, which offices will come to me presently."

In appearance Bentley was tall and powerful, the forehead was high and not very broad, but the great development and rather coarse lines of the lower part of the face and cheeks seem to me the most prominent features and always strike me as indicative of cruelty and selfishness. The hair was brown and the hands small. Of his appearance Prof. Jebb says, "The pose of the head is haughty, almost defiant; the eyes, which are large, prominent, and full of bold vivacity, have a light in them as if Bentley were looking straight at an impostor whom he detected, but who still amused him; the nose, strong and slightly tip-tilted, is moulded as if nature had wished to shew what a nose can do for the combined expression of scorn and sagacity; and the general effect of the countenance, at a first glance, is one which suggests power—frank, self-assured, sarcastic, and I fear we must add insolent."

In character he was warm-hearted, impulsive, and no doubt well-intentioned; and separated from him by a century and a half we may give him credit for the reforms he made—in spite of the illegal manner in which they were introduced, and of his injustice and petty meanness against those who opposed him. Even his apologists admit that he was grasping, arrogant, arbitrary, intolerant, and at any rate in manner not a gentleman, while in the latter part of his life he neglected the duties of his office. But his abilities immeasurably exceeded those of his contemporaries, and such as he was he has left a permanent impress on the history of Cambridge.

The interest that Bentley felt in the Newtonian philosophy arose from the nature of the conclusions and of the irrefutable logic by which they were proved. He was not however capable of appreciating the mathematical analysis by which they had been attained. Of those who were urged by him to take up the study of mathematics, one of the earliest was Whiston. **William Whiston**[1] was born in Leicestershire on Dec. 9, 1667. He entered in 1685 at Clare, and mentions in his biography that he attended Newton's lectures. He took his B.A. in the Lent term of 1690, in the same year was elected a fellow, and for some time subsequently took pupils. In 1696 he published his celebrated *Theory of the earth.* The fanciful manner in which he accounted for the deluge by means of the tail of a comet is well known; but Bentley's criticism that Whiston had forgotten to provide any means for getting rid of the water with which he had covered the earth, and that it was of little use to explain the origin of the deluge by natural means if it were necessary to invoke the aid of the Almighty to finish the operation, is a sound one.

When in 1699 Newton was appointed master of the mint he asked Whiston to act as his deputy in the Lucasian chair. As such Whiston lectured on the *Principia.* In 1703 Newton resigned his professorship and Whiston was chosen as his successor.

In 1702 Whiston brought out an edition of Tacquet's[2]

[1] Whiston wrote an autobiography, published at London in 1749, but many of the events related are not described accurately: see Monk's *Life of Bentley*, vol. i. pp. 133, 151, 215, 290, and vol. ii. p. 18. An account of his life is given in the *Biographia Britannica*, first edition, 6 vols., London, 1747—66.

[2] *Andrew Tacquet*, who was born at Antwerp in 1611 and died in 1660, was one of the best known Jesuit mathematicians and teachers of the seventeenth century. His translation of Euclid's *Elements* was published in 1655, and remained a standard text-book on the continent until superseded by Legendre's *Géometrie.* Tacquet also wrote on optics and astronomy. His collected works were republished in two volumes at Antwerp in 1669.

Euclid which remained the standard English text-book on elementary geometry until displaced by the edition of Robert Simson issued in 1756. A year or so later Whiston asked Newton to be allowed to print the *Universal arithmetic*, manuscript copies of which were circulating in the university in much the same way as manuscripts containing matter which has not yet got incorporated into text-books do at the present time. Newton gave a reluctant consent, and it was published by Whiston in 1707.

Whiston seems to have been an honest and well-meaning man but narrow, dogmatic, and intolerant; and having adopted certain religious opinions he not only preached them on all occasions, but he questioned the honesty of those who differed from him. The following account of the beginning of the controversy is taken from a letter of William Reneu of Jesus, an undergraduate of the time.

> I have a peice of very ill news to send you i.e. viz. y[t] one *Whiston* our Mathematicall Professor, a very learned (and as we thought pious) man has written a Book concerning y[e] Trinity and designs to print it, wherein he sides wth y[e] Arrians; he has showed it to severall of his freinds, who tell him it is a damnable, heretical Book and that, if he prints it, he'll Lose his Professorship, be suspended ab officio et beneficio, but all won't do, he saies, he can't satisfy his Conscience, unless he informs y[e] world better as he thinks than it is at present, concerning y[e] Trinity.

It is characteristic of the tolerancy of the Cambridge of the time that, although Whiston's opinions were contrary to the oath he had taken on commencing his M.A., yet no public notice was taken of them until he began to attack individuals who did not agree with him. It was impossible to allow the scandal thus occasioned to continue indefinitely. Whiston was warned and as he persisted in going on he was in 1711 expelled from his chair. The details of his opinions are now of no interest.

After leaving the university Whiston wrote several books on astronomy and theology, but they are not material to my purpose. A list of them will be found in his life. His translation of Josephus is still in common use. He and Desaguliers gave lectures on experimental physics illustrated by experiments in or about 1714: these are said to have been the earliest of the kind delivered in London.

An attempt to prosecute him was made in London by some clergymen; but the courts deemed it vindictive, and strained the law to delay the sentence till 1715, when all past heresy was pardoned by an act of grace. Whiston rather cleverly made use of these proceedings to push his opinions and in particular his theory of the deluge into general notice: on one occasion he put an account of the latter instead of a petition into the legal pleadings and the judges discussed it with great gravity and bewilderment until they found it had nothing to do with the suit. As so often happened in similar cases the prosecution only served to disseminate his opinions and excite sympathy for his undoubted honesty and candour. Queen Caroline who liked to see celebrated heretics ordered him to preach before her, and after the sermon in talking to him said she wished he would tell her of any faults in her character, to which he replied that talking in public worship was certainly a prominent one, and on her asking whether there were any others he refused to tell her till she had amended that one. He died in London on Aug. 22, 1752.

Intolerant, narrow, vain, and with no idea of social proprieties[1] he was yet honest and courageous; and though not a specially distinguished mathematician himself, his services in disseminating the discoveries of others were considerable. His tenure of the professorship was marked by the publication of Newton's writings on algebra and theory of equations (the *Universal arithmetic*), analytical geometry (cubic curves), the fluxional calculus, and optics. Copies of lectures and papers in the transactions of learned societies are and always will be inaccessible to many students. Henceforth Newton's mathematical works were open to all readers, and the credit of that is partly due to Whiston.

[1] See e.g. p. 183 of his memoirs.

Whiston was succeeded in the Lucasian chair by Saunderson. **Nicholas Saunderson**[1] was born in Yorkshire in 1682, and became blind a few months after his birth. Nevertheless he acquired considerable proficiency in mathematics, and was also a good classical scholar. When he grew up he determined to make an effort to support himself by teaching, and attracted by the growing reputation of the Cambridge school he moved to Cambridge, residing in Christ's College. There with the permission of Whiston he gave lectures on the *Universal arithmetic*, *Optics*, and *Principia* of Newton, and drew considerable audiences. His blindness, poverty, and zeal for the study of mathematics procured him many friends and pupils; and among the former are to be reckoned Newton and Whiston.

When in 1711 Whiston was expelled from the Lucasian chair, queen Anne conferred the degree of M.A. by special patent on Saunderson so as to qualify him to hold that professorship, and he continued to occupy it till his death on April 19, 1739.

His lectures on algebra and fluxions were embodied in text-books published posthumously in 1740 and 1756. The algebra contains a description of the board and pegs by the use of which he was enabled to represent numbers and perform numerical calculations. The work on fluxions contains his illustrations of the *Principia* and of Cotes's *Logometria*; and probably gives a fair idea of how the subject was treated in the Cambridge lecture-rooms of the time.

He is described by one of his pupils as "justly famous not only for the display he made of the several methods of reasoning, for the improvement of the mind, and the application of mathematics to natural philosophy; but by the reverential regard for Truth as the great law of the God of truth, with which he endeavoured to inspire his scholars, and that peculiar felicity in teaching whereby he made his subject familiar to

[1] An account of his life is prefixed to his *Algebra* published in two volumes at Cambridge in 1740.

their minds." He was passionate, outspoken, and truthful, and seems to be fairly described as "better qualified to inspire admiration than to make or preserve friends."

I notice references to two other mathematicians of this time as having taken a prominent part in the introduction of the Newtonian philosophy, but I can find no particulars of their lives or works. The first of these is **Thomas Byrdall**, of King's College, who died in 1721, and is said to have not only assisted Newton in preparing the *Principia* for the press, but to have checked most of the numerical calculations. Contemporary rumour is not to be lightly rejected, but I have never seen any evidence for the statement. The second of these writers is **James Jurin**, a fellow of Trinity College, who was born in 1684, graduated as B.A. in 1705, and died in 1750. He wrote in 1732 on the theory of vision, and was one of the earliest philosophers who tried to apply mathematics to physiology. He took a prominent part in the controversies between the followers of Newton and Leibnitz, and in particular engaged in a long dispute[1] with Michelotti on a question connected with the momentum of running water.

During this time the Newtonian philosophy had become dominant in the mathematical schools at Oxford: the Savilian professors of astronomy being David Gregory from 1691 to 1708, and John Keill from 1708 to 1721; and the Savilian professors of geometry being Wallis (see p. 42) till 1703, and thence till 1720 Edmund Halley; but mathematics was still an exotic study there, and the majority of the residents regarded mathematics and puritanism as allied and equally unholy subjects. In London the Newtonian philosophy was worthily represented by Abraham de Moivre and by Brook Taylor, while Newton himself regularly presided at the meetings of the Royal Society.

[1] See *Philosophical transactions* vols. LX. to LXVI.

The only one of those immediately above mentioned who came from Cambridge was **Brook Taylor**[1], who was born at Edmonton on Aug. 18, 1685, and died in London on Dec. 29, 1731. He entered at St John's College in 1705, and graduated as LL.B. in 1709. After taking his degree he went to live in London, and from the year 1708 onwards he wrote numerous papers in the *Philosophical transactions*, in which among other things he discussed the motion of projectiles, the centre of oscillation, and the forms of liquids raised by capillarity. He wrote on linear perspective, two volumes, 1715 and 1719. But the work by which he is generally known is his *Methodus incrementorum directa et inversa* published in 1715. This contained the enunciation and a proof of the well-known theorem

$$f(x+h) = f(x) + hf'(x) + \frac{h^2}{\underline{|2}} f''(x) + \ldots,$$

by which any function of a single variable can be expanded. He did not consider the convergency of the series, and the proof, which contains numerous assumptions, is not worth reproducing. In this treatise he also applied the calculus to various physical problems, and in particular to the theory of the transverse vibrations of strings.

Regarded as mathematicians, Whiston, Laughton, and Saunderson barely escape mediocrity, but their contemporary Cotes, of whom I have next to speak, was a mathematician of exceptional power, and his early death was a serious blow to the Cambridge school. The remark of Newton that if only Cotes had lived "we should have learnt something" indicates the opinion of his abilities generally held by his contemporaries.

Roger Cotes[2] was born near Leicester on July 10, 1682. He entered at Trinity in 1699, took his B.A. in 1703, and in

[1] An account of his life by Sir William Young is prefixed to the *Contemplatio philosophica*, London, 1793.

[2] See the *Biographia Britannica*, second edition, London, 1778—93, and also the *Dictionary of national biography*.

1705 was elected to a fellowship. In 1704 Dr Plume, the archdeacon of Rochester and formerly of Christ's College (bachelor of theology, 1661), founded a chair of astronomy and experimental philosophy. The first appointment was made in 1707, and Cotes was elected[1]. Whiston was one of the electors, and he writes, "I was the only professor of mathematics directly concerned in the choice, so my determination naturally had its weight among the rest of the electors. I said that I pretended myself to be not much inferior in mathematics to the other candidate's master, Dr Harris, but confessed that I was but a child to Mr Cotes: so the votes were unanimous for him[2]." Newton, to whom Bentley had introduced Cotes, also wrote a very strong testimonial in his favour.

Bentley at once urged the new professor to establish an astronomical observatory in the university. The university gave no assistance, but Trinity College consented to have one erected on the top of the Great Gate, and to allow the Plumian professor to occupy the rooms in connection with it; considerable subscriptions were also raised in the college to provide apparatus. The observatory was pulled down in 1797.

In 1709 Newton was persuaded to allow Cotes to prepare the long-talked-of second edition of the *Principia.* The first edition had been out of print by 1690; but though Newton had collected some materials for a second and enlarged edition, he could not at first obtain the requisite data from Flamsteed, the astronomer-royal, and subsequently he was unable or unwilling to find the time for the necessary revision. The second edition was issued in March 1713, but a considerable part of the

[1] The successive professors were as follows. From 1707 to 1716, Roger Cotes of Trinity; from 1716 to 1760, Robert Smith of Trinity (see p. 91); from 1760 to 1796, Anthony Shepherd of Christ's (see p. 103); from 1796 to 1822, Samuel Vince of Caius (see p. 103); from 1822 to 1828, Robert Woodhouse of Caius (see p. 118); from 1828 to 1836, Sir George B. Airy of Trinity (see p. 132); from 1836 to 1883, James Challis of Trinity (see p. 132); who in 1883 was succeeded by G. H. Darwin of Trinity, the present professor.

[2] See p. 133 of Whiston's *Memoirs.*

new work contained in it was due to Cotes and not to Newton. The whole correspondence between Newton and Cotes on the various alterations made in this edition is preserved in the library of Trinity College, Cambridge: it was edited by Edleston for the college in 1850. This edition was sold out within a few months, but a reproduction published at Amsterdam supplied the demand. Cotes himself died on June 5, 1716, shortly after the completion of this work.

He is described as possessing an amiable disposition, an imperturbable temper, and a striking presence; and he was certainly loved and regretted by all who knew him.

His writings were collected and published in 1722 under the titles *Harmonia mensurarum* and *Opera miscellanea*. His professorial lectures on hydrostatics were published in 1738. A large part of the *Harmonia mensurarum* is given up to the decomposition and integration of rational algebraical expressions; that part which deals with the theory of partial fractions was left unfinished, but was completed by de Moivre. Cotes's theorem in trigonometry which depends on forming the quadratic factors of $x^n - 1$ is well known. The proposition that "if from a fixed point O a line be drawn cutting a curve in Q_1, Q_2.. Q_n, and a point P be taken on it so that the reciprocal of OP is the arithmetic mean of the reciprocals of OQ_1, OQ_2,...OQ_n, then the locus of P will be a straight line" is also due to Cotes. The title of the book was derived from the latter theorem. The *Opera miscellanea* contains a paper on the method for determining the most probable result from a number of observations: this was the earliest attempt to frame a theory of errors. It also contains essays on Newton's *Methodus differentialis*, on the construction of tables by the method of differences, on the descent of a body under gravity, on the cycloidal pendulum, and on projectiles.

It was unfortunate for Cotes's reputation that his friend Brook Taylor stated the property of the circle which Cotes had discovered as a challenge to foreign mathematicians in a manner which was somewhat offensive. John Bernoulli solved the question proposed in 1719, and his friends seized on his triumph as a convenient opportunity for shewing their dislike of Newton by depreciating Cotes.

The study of mathematics in the different colleges received at this time a considerable stimulus by the establishment in 1710 of certain lectureships by Lady Sadler. On the advice of William Croone (born about 1629 and died in 1684), a fellow of Emmanuel and professor of rhetoric at Gresham College, she gave to the university an estate of which the income was to be divided amongst the lecturers on algebra at certain colleges. This no doubt helped to promote the interest in that subject during the seventeenth century. With the advance in the standard of education it ceased to be productive of much benefit, and in 1860 it was changed into a professorship of pure mathematics; in 1863 Arthur Cayley of Trinity was appointed professor.

Cotes was succeeded as Plumian professor by his cousin Robert Smith. **Robert Smith** was born in 1689, entered at Trinity in 1707, took his B.A. in 1711, and was elected to a fellowship in the following year. He held the office of master of mechanics to the king. As Plumian professor he lectured on optics and hydrostatics, and subsequently he wrote text-books on both those subjects. His *Opticks* published in 1728 is one of the best text-books on the subject that has yet appeared, and with a few additions might be usefully reprinted now. He also published in 1744 a work on sound, entitled *Harmonics*, which contains the substance of lectures he had for many years been giving. He edited Cotes's works. He was made master of Trinity in 1742, and died at Cambridge on Feb. 2, 1768. He founded by his will two annual prizes for proficiency in mathematics and natural philosophy, to be held by commencing bachelors and known by his name. They proved productive of the best results, and at a later time they enabled the university to encourage some of the higher branches of mathematics which did not directly come into the university examinations for degrees.

The labours of Laughton, Bentley, Whiston, Saunderson, Cotes, and Smith were rewarded by the definite establishment about the year 1730 of the Newtonian philosophy in the schools of the university. The earliest appearance of that philosophy in the scholastic exercises is the act kept by Samuel Clarke in 1694 and above alluded to. Ten years later it was not unusual to keep one act from Newton's writings; but from 1730 onwards it was customary to require at least one disputation to be on a mathematical subject—usually on Newton—and in general to expect one to be on a philosophical thesis, although after 1750 it was possible to propose mathematical questions only. The decade from 1725 to 1735 is an important one in a history of mathematics at Cambridge, not only for the reasons given above, but because the mathematical tripos, which profoundly affected the subsequent development of mathematics in the university, originated then. The history of the origin and growth of that examination may be left for the present. The death of Newton and the retirement or death of nearly all those who had been brought under his direct influence also fall within this decade, and it thus naturally marks the conclusion of this chapter.

The effect of the teaching of the above-mentioned mathematicians in extending the range of reading is shewn by the following list of mathematical text-books which were in common use by the year 1730. The dates given are those of the first editions, but in most cases later editions had been issued incorporating the discoveries of subsequent writers.

First, for the subjects of pure mathematics. The usual text-books on pure geometry were the *Elements* of Euclid (editions of Barrow, Gregory, or Whiston), the *Conics* of Apollonius (Halley's edition, 1710), or of de Lahire (1685), to which we may perhaps add the fourth and fifth sections of the first book of the *Principia*. [Simson's *Conics* was published in 1735, and became the recognized text-book for that subject for the remainder of the eighteenth century.] The usual text-book on arithmetic was Oughtred's *Clavis*, or E. Wingate's *Arithmetic* (1630). The usual text-books on algebra were those by Harriot, Oughtred, Wallis, and Newton (*Universal arithmetic*). The usual text-books on trigonometry were those by Oughtred (the *Clavis*), Seth Ward (1654), Caswell (1685), and E. Wells (1714). The usual text-books on analytical geometry were those by Wallis (1665), and Maclaurin (1720). The usual text-books on the infinitesimal calculus were those by Humphry Ditton (1704), W. Jones (1711), and Brook Taylor (1715).

Next for the subjects of applied mathematics. I know of no work on mechanics of this time suitable for students other than the treatises by Stevinus, Huygens, and Wallis, and the introduction to the *Principia*: no one of these is what we should call a text-book.

Geometrical optics was generally studied in the pages of Newton, Gregory (1695), or Robert Smith (1728). In elementary hydrostatics a translation of a text-book by Mariotte was used, but copies or notes of the lectures of Cotes and Whiston were probably accessible. The elements of both the last-named and other physical subjects were also read in W. J. 'sGravesande's work (published in 1720 and translated by Desaguliers in 1738). The mathematical treatment of the higher parts of the subject, if studied at all, was read in the edition of Newton's lectures.

There were numerous works on astronomy in common use. Selected portions of the *Principia*, Clarke's translation and commentary on Rohault, and Kepler's writings were read by the more advanced students, but I suspect that most men contented themselves with one or more of the popular summaries of which several were then in circulation—one of the best being that by David Gregory (1702).

Of course a much longer list of text-books then obtainable might be drawn up, but I think the above includes all, or nearly all, the books then in common use. I believe the writings of Leibnitz, the Bernoullis, and their immediate followers were but rarely consulted, though they probably were included in the more important mathematical libraries of the time. I may here add that the libraries of Cotes and Robert Smith are both preserved in Trinity.

Two tutors of a somewhat earlier date drew out time tables shewing the order in which the subjects should be read, accompanied by a list of the books in common use. They are published in the third and fourth appendices to the *Scholae academicae*, from which the following account is condensed.

In the *Student's guide* written about 1706 by Daniel Waterland, a fellow and subsequently master of Magdalene College, the following course of reading in "philosophical studies" is recommended: Waterland adds that by January and February he means the two first months of residence and not necessarily the calendar months named. It will be noticed

	First year	*Second year*	*Third year*	*Fourth year*
Jan. Feb.	Wells's Arithm.	Wells's Astron. Locke.	Burnet's Theory with Keill's Remarks.	Baronius's Metaphysicks.
March April	Euclid's Elem.	Locke's Hum. Und. De la Hire Con. Sect.	Whiston's Theory with Keill's Remarks.	Newton's Opticks.
May June	Euclid's Elem. Burgersdicius's Logick.	Whiston's Astron.	Wells's Chron. Beveridge's Chronology.	Whiston's Praelect. Phys. Math.
July Aug.	Euclid's Elem. Burgersdicius.	Keil's Introduction.	Whitby's Eth. Puffendorf's Law of Nat.	Gregory's Astronomy.
Sept. Oct.	Wells's Geogr.	Cheyne's Phil. Principles.	Puffendorf. Grotius de Jure Belli.	
Nov. Dec.	Wells's Trig. Newton's Trig.	Rohault's Physics.	Puffendorf. Grotius.	

that a mathematician was expected to read the elements of various sciences, and the curriculum was not a narrow one.

Waterland remarks on this course that Hammond's *Algebra*, Wells's *Mechanics*, and Wells's *Optics* should also be added at some time in the first three years. Further, a bachelor if he did not intend to take orders should before proceeding to the M.A. degree read Newton's *Principia*, Ozanam's *Cursus*, Sturmius's *Works*, Huygens's *Works*, Newton's *Algebra*, and Milnes's *Conic sections*.

In a third edition issued in 1740 the *Arithmetic*, *Trigonometry*, and *Astronomy* of Wells are respectively replaced by Wingate's *Arithmetic*, Keill's *Trigonometry*, and Harris's *Astronomy*; Simpson's *Conics* is substituted for that by de la Hire; Bartholin's *Physics* is to be read as well as Rohault's; finally Whiston's *Astronomy* is struck out and Milnes's *Conic sections* recommended to be then read. Besides these the attention of the student is directed to Maclaurin's *Algebra*, Simpson's *Algebra*, and Huygens's *Planetary worlds*.

A somewhat similar course was sketched out in 1707 by Robert Green, a fellow and tutor of Clare, who took his B.A. in 1699 and died in 1730. Green was almost the last Cantab of any position who rejected the Newtonian theory of physical astronomy. He recommended his pupils to spend the first year on the study of *classics*: the second on *logic*, *ethics*, *geometry* (Euclid, Sturmius, Pardies, or Jones), *arithmetic* (Wells, Tacquet, or Jones), *algebra* (Pell, Wallis, Harriot, Kersey, Newton, Descartes, Harris, Oughtred, Ward, or Jones), and *corpuscular philosophy* (Descartes, Rohault, Varenius, Le Clerk, or Boyle): the third on *natural science*, *optics* (Gregory, Rohault, Dechales, Barrow, NEWTON, Descartes, Huygens, Kepler, or Molyneux), and *conic sections and other curves* (De Witt, De Lahire, Sturmius, L'Hospital, Newton, Milnes, or Wallis): the fourth year on *mechanics* of solids and fluids (Marriotte, Keill, Huygens, Sturmius, Boyle, Newton, Ditton, Wallis, Borellus, or Halley), *fluxions and infinite series* (Wallis, Newton, Raphson, Hays, DITTON, Jones, Nieuwentius, or

L'Hospital), *astronomy* (Gassendi, Mercator, BULLIALDUS, Horrocks, Flamsteed, Newton, Gregory, Whiston, or Kepler), and *logarithms* and *trigonometry* (Sturmius, Briggs, Vlacq, Gellibrand, Harris, Mercator, Jones, Newton, or Caswell). The authors whose names are printed in small capitals are those specially recommended. The order in which the subjects are to be taken is curious.

CHAPTER VI.

THE LATER NEWTONIAN SCHOOL.

CIRC. 1730—1820.

I HAVE already explained that the results of the infinitesimal calculus may be expressed in either of two notations. In most modern books both are used, but if we must confine ourselves to one then that adopted by Leibnitz is superior to that used by Newton, and for some applications—such as the calculus of variations—is almost essential. The question as to the relative merits of the two methods was unfortunately mixed up with the question as to whether Leibnitz had discovered the fundamental ideas of the calculus for himself, or whether he had acquired them from Newton's papers, some of which date back to 1666. Personal feelings and even national jealousies were appealed to by both sides. Finally Newton's notation was generally adopted in England, while that invented by Leibnitz was employed by most continental mathematicians. The latter result was largely due to the influence of John Bernoulli, the most famous and successful mathematical teacher of his age, who through his pupils (especially Euler) determined the lines on which mathematics was developed on the continent during the larger part of the eighteenth century.

A common language and facility of intercommunication of ideas are of the utmost importance in science, and even if the Cambridge school had enjoyed the use of a better notation than their continental contemporaries they would have lost a great deal by their isolation. So little however did they realize this truth that they made no serious efforts to keep themselves acquainted with the development of analysis by their neighbours. On the continent on the other hand the results arrived at by Newton, Taylor, Maclaurin, and others were translated from the fluxional into the differential notation almost as soon as they were published; to this I should add that the journals and transactions in which continental mathematicians embodied their discoveries were circulated over a very wide area and large numbers of them were distributed gratuitously.

The use of the differential notation may be taken as definitely adopted on the continent about the year 1730. The separation of the Newtonian school from the general stream of European thought begins to be observable about that time, and explains why I closed the last chapter at that date.

Modern analysis is derived from the writings of Leibnitz and John Bernoulli as interpreted by d'Alembert, Euler, Lagrange, and Laplace. Even to the end the English school of the latter half of the eighteenth century never brought itself into touch with these writers. Its history therefore leads nowhere, and hence it is not necessary to discuss it at any great length.

The isolation of the later Newtonian school would sufficiently account for the rapid falling off in the quality of the work produced, but the effect was intensified by the manner in which its members confined themselves to geometrical demonstrations. If Newton had given geometrical proofs of most of the theorems in the *Principia* it was because their validity was unimpeachable, and as his results were opposed to the views then prevalent he did not wish the discussion as to their truth to turn on the correctness of the methods used to demonstrate them. But his followers, long after the principles of the infinitesimal calculus had been universally recognized as valid, continued to employ geometrical proofs wherever it was possible. These proofs are elegant and ingenious, but it is necessary to find a separate kind of demonstration for every distinct class of problems so that the processes are not nearly so general as those of analysis.

During the whole of the period treated in this chapter only two mathematicians of the first rank can be claimed for the Newtonian school. These were Maclaurin in Scotland and Clairaut in France: the latter being the sole distinguished foreigner who by choice used the Newtonian geometrical methods. Neither of them had any special connection with Cambridge. Waring might perhaps under more favourable circumstances have taken equal rank with them, but except for him I can recall the names of no Cambridge men whose writings at this distance of time are worth more than a passing notice.

Although the quality of the mathematical work produced in this period was so mediocre yet the number of eminent lawyers educated in the mathematical schools of Cambridge was extraordinarily large. Many careful observers have asserted that in the majority of cases a mathematical training affords the ideal general education which a lawyer should have before he begins to read law itself. A study of analytical mathematics is among the best instruments for training the reasoning faculties, and for many students it provides the best available preliminary education for a scientific lawyer; but I doubt if it has that special fitness which geometry and the use of geometrical methods seem to possess for the purpose.

Throughout the time considered in this chapter the Newtonian philosophy was dominant in the schools of the university, but the senate-house examination gradually took the place of the scholastic exercises as the real test of a man's abilities. An account of those exercises and of the origin and development of the mathematical tripos is given in chapters IX. and X. I will merely here remark that the tripos (then known as the senate-house examination) became by the middle of the eighteenth century the only avenue to a degree, and that all undergraduates from that time forward had to read at least the elements of mathematics.

Of course geometry, algebra, and the fluxional calculus were read by all mathematical students; but the subjects which attracted most attention during this time were astronomy and optics. The papers in the transactions of the Royal Society and the problems published in the form of challenges in the pages of the *Ladies' diary* (1707—1817) and other similar publications will give a fair idea of the kind of questions that excited most interest in England. If any one will compare these with the papers then being published on the continent by d'Alembert, Euler, Lagrange, Laplace, Legendre, Gauss, and others he will not I think blame me for making my account of the Cambridge mathematical school of this time little else than a list of names.

I shall first consider very briefly the mathematical professors of this time, and shall then similarly enumerate a few other contemporary mathematicians and physicists.

I begin then by mentioning the professors.

The occupants of the Lucasian chair were successively John Colson, Edward Waring, and Isaac Milner. Saunderson died in 1739, and was succeeded by Colson. **John Colson**[1] was born at Lichfield in 1680. In 1707 he communicated a paper to the Royal Society on the solution of cubic and biquadratic equations. He was then a schoolmaster, and having acquired some reputation as a successful teacher was recommended by Robert Smith the master of Trinity to come to Cambridge and lecture there. He had rooms in Sidney, but apparently was not a member of that college: subsequently he moved to Emmanuel, whence he took his M.A. degree in 1728. While residing there he contributed a paper on the principles of algebra to the *Philosophical transactions*, 1726.

He then accepted a mastership at Rochester grammar-

[1] No contemporary biography of Colson is extant; but nearly all the known references to him have been collected in the *Dictionary of national biography*.

school. In 1735 he wrote a paper on spherical maps[1]; and in 1736 he published the original manuscript of Newton on fluxions, together with a commentary (see pp. 70, 71).

When a candidate for the Lucasian chair in 1739 he was opposed by Abraham de Moivre, who was admitted a member of Trinity College and created M.A. to qualify him for the office. Smith really decided the election, and as de Moivre was very old and almost in his dotage he pressed the claims of Colson. The appointment was admitted to be a mistake, and even Cole, who was a warm friend of Colson, remarks that the latter merely turned out to be "a plain honest man of great industry and assiduity, but the university was much disappointed in its expectations of a professor that was to give credit to it by his lectures." Colson died at Cambridge on Jan. 20, 1760.

Besides the papers sent to the Royal Society enumerated above and his edition of Newton's *Fluxions*, Colson wrote an introductory essay to Saunderson's *Algebra*, 1740, and made a translation of Agnesi's treatise on analysis: he completed the latter just before his death, and it was published by baron Maseres in 1801.

Colson was succeeded in 1760 by Waring, a fellow of Magdalene. **Edward Waring** was born near Shrewsbury in 1736, took his B.A. as senior wrangler in 1757, and died on Aug. 15, 1798. He is described as being a man of unimpeachable honour and uprightness but painfully shy and diffident. The rival candidate for the Lucasian chair was Maseres; and as Waring was not then of standing to take the M.A. degree he had to get a special license from the crown to hold the professorship.

Waring wrote *Miscellanea analytica*, issued in 1762, *Meditationes algebraicae*, issued in 1770, *Proprietates algebraicarum curvarum*, issued in 1772; and *Meditationes analyticae*, issued in 1776. The first of these is on algebra and analytical geometry,

[1] *Philosophical transactions* 1735.

and includes some papers published when he was a candidate for the Lucasian chair as a proof of his fitness for the post. The third of these works is that which is most celebrated: it contains several results that were previously unknown. From a cursory inspection of these writings I think they shew considerable power, but the classification and arrangement of them are imperfect.

Waring contributed numerous papers to the *Philosophical transactions*. Most of these are on the summation of series, but in one of them, read in 1778, he enunciated a general method for the solution of an algebraical equation which is still sometimes inserted in text-books; his rule is correct in principle but involves the solution of a subsidiary equation which is sometimes of a higher order than the equation originally proposed. Papers by him on various algebraical problems will be found in the *Philosophical transactions* for 1763, 1764, 1779, 1784, 1786, 1787, 1788, 1789, and 1791.

In a reply to some criticisms which had been made on the first of the above-mentioned works he enunciated the celebrated theorem that if p be a prime then $1 + |\underline{p-1}$ is a multiple of p; for this result he was indebted to one of his pupils, *John Wilson*, who was then an undergraduate at Peterhouse. Wilson was born in Cumberland on Aug. 6, 1741, graduated as senior wrangler in 1761, and subsequently took pupils. He was a good teacher and made his pupils work hard, but sometimes when they came for their lessons they found the door sported and 'gone a fishing' written on the outside, which Paley (who was one of them) deemed the addition of insult to injury, for he was himself very fond of that sport. Wilson later went to the bar, and was appointed a justice in the Common Pleas. He died at Kendal on Oct. 18, 1793.

Waring was succeeded in 1798 by Milner, who was then professor of natural philosophy, master of Queens' College, and dean of Carlisle. **Isaac Milner**[1] was born at Leeds in

[1] His life has been written by Mary Milner, London, 1842.

1751, took his B.A. in 1774 as senior wrangler, and died in London on April 1, 1820. He wrote several works on theology. A contemporary says that he had "extensive learning always at his command..... great talents for conversation and a dignified simplicity of manner," but he does not seem to have possessed any special qualifications for the Lucasian chair. At an earlier time he had frequently taken part in the examinations in the senate-house, but I believe I am right in saying that after his election to the professorship he never lectured, or taught, or examined in the tripos, or presided in the schools.

The occupants of the Plumian chair during the period treated in this chapter were Robert Smith (see p. 91), Anthony Shepherd, and Samuel Vince.

In 1760 Robert Smith was succeeded by Shepherd. **Anthony Shepherd** was born in Westmoreland in 1722, took his B.A. from St John's in 1743, was subsequently elected a fellow of Christ's, and died in London on June 15, 1795. Of him I know nothing save that in 1772 he published some refraction and parallax tables, and that in 1776 he printed a list of some experiments on natural philosophy which he had used to illustrate a course of lectures he had given in Trinity College.

Shepherd was followed in 1796 by Vince, a fellow of Caius. **Samuel Vince** was born in Suffolk about 1754, took his B.A. as senior wrangler in 1775, and died in December, 1821. His original researches consisted chiefly of numerous observations on the laws of friction and the motion of fluids, and he contributed papers on these subjects to the *Philosophical transactions* for 1785, 1795, and 1798. His results are substantially correct. A list of all his papers sent to various societies is given in Poggendorff. His most important work is an astronomy published in three volumes at Cambridge, 1797—1808; the first volume is descriptive, the second an account of physical astronomy, and the third a collection of tables arranged for

English observers: this was preceded by a work on practical astronomy issued in 1790.

He also wrote text-books on conic sections, algebra, trigonometry, fluxions, the lever, hydrostatics, and gravitation, which form part of a general course of mathematics: these were all published or reissued in 1805 or 1806, and for a short time were recognised as standard text-books for the tripos; but they are badly arranged and were superseded by the works of Wood. His treatise on fluxions first published in 1805 went through numerous editions, and is one of the best expositions of that method. In it, however, as in all the Cambridge works of that time, he used $\dot{x}$ to denote, not the fluxion of x, but the increment of x generated in a small time; that is what Newton would have written as $\dot{x}o$. He asserts that "this is agreeable to Sir I. Newton's ideas on the subject," and "as the velocities are in proportion to the increments or decrements which would be generated in a given time, if at any instant the velocities were to become uniform, such increments or decrements will represent the fluxions at that instant[1]." He also used the symbol of integration (see p. 71).

A public advertisement of his lectures for 1802 is as follows.

> The lectures are experimental, comprising mechanics, hydrostatics, optics, astronomy, magnetism, and electricity; and are adapted to the plan usually followed by the tutors in the university. All the fundamental propositions in the first four branches, are proved by experiments, and accompanied with such explanations as may be useful to the theoretical student. Various machines and philosophical instruments are exhibited in the course of the lectures, and their construction and use explained. And in the two latter branches a set of experiments are instituted to shew all the various phenomena, and such as tend to illustrate the different theories which have been invented to account for them. The lectures are always given in the first half of the midsummer term at 4 o'clock in the afternoon, in the public Lecture-room under the front of the Public Library. Terms are 3 guineas for the first course, 2 guineas for the second, and afterwards gratis.

[1] Vince's *Fluxions*, p. 1.

A "plan" of his lectures with a detailed account of his experiments was published in 1793, and another one was issued in 1797. His lectures are said to have been good, and I believe he was always willing to assist students in their reading. His successors will be mentioned in the next chapter.

In 1749 Thomas Lowndes of Overton founded another professorship[1] of astronomy and geometry. The first occupant of the chair was **Roger Long**, a fellow and subsequently master of Pembroke College, and the friend of the poet Gray. Long was born in Norfolk on Feb. 2, 1680, graduated as B.A. in 1701, and died on Dec. 16, 1770. His chief work is one on astronomy in two quarto volumes published in 1742: fresh editions were issued in 1764 and 1784, and it became a standard text-book at Cambridge; the descriptive parts are said to be well written. In 1765, or according to some accounts 1753, he constructed a zodiack or large sphere capable of containing several people and on the inside of which the constellations visible from Cambridge were marked. This famous globe stood in the grounds of Pembroke College, and was only destroyed in 1871.

Long was succeeded in 1771 by **John Smith**, the master of Caius College, who in his turn was followed in 1795 by **William Lax**, a fellow of Trinity, who was born in 1751 and held the chair till his death on Oct. 29, 1836. Both of these professors seem to have neither lectured nor taught. Lax wrote a pamphlet on Euclid, 1808: and in 1821 issued some tables for use with the *Nautical almanack*. He also contributed papers to the *Philosophical transactions* for 1799 and 1809.

[1] The successive professors were as follows. From 1749 to 1771, Roger Long of Pembroke; from 1771 to 1795, John Smith of Caius; from 1795 to 1836, William Lax of Trinity; from 1836 to 1858, George Peacock of Trinity (see p. 124); who in 1858 was succeeded by J. C. Adams of Pembroke, the present professor.

To meet the want of the lectures they should have given **Francis John Hyde Wollaston** (born about 1761, took his B.A. in 1783, and died in 1823), a fellow of Trinity Hall and Jacksonian professor, lectured on astronomy from 1785 to 1795, and **William Farish** (born in 1759 and died in 1837), a fellow of Magdalene, who was professor of chemistry from 1794 to 1813 and of natural experimental philosophy from 1813 to 1837, lectured on mechanics. A paper by Farish on isometrical perspective appears in the *Cambridge philosophical transactions* for 1822.

Farish was also vicar of St Giles's, Cambridge, and many stories of the complications produced by his extraordinary absence of mind are still current. He is celebrated in the domestic history of the university for having reduced the practice of using Latin as the official language of the schools and the university to a complete farce. On one occasion, when the audience in the schools was unexpectedly increased by the presence of a dog, he stopped the discussion to give the peremptory order *Verte canem ex.* At another time one of the candidates had forgotten to put on the bands which are still worn on certain ceremonial occasions. Farish, who was presiding, said, *Domine opponentium tertie, non habes quod debes. Ubi sunt tui*...(with a long pause) *Anglice bands?* To whom with commendable promptness the undergraduate replied, *Dignissime domine moderator, sunt in meo* (*Anglice*) *pocket.* Another piece of scholastic Latin quoted by Wordsworth is, *Domine opponens non video vim tuum argumentum*[1].

The only other mathematicians of this time whom I deem it necessary to mention here are George Atwood, Miles Bland, Bewick Bridge, John Brinkley, Daniel Cresswell, William Frend, Francis Maseres, Nevil Maskelyne, John Rowning, Francis Wollaston, and James Wood. I confine myself to a

[1] See p. 41 of the *Scholae academicae*; and Nichol's *Literary anecdotes*, VIII. 541.

short note on each, and I have arranged these notes roughly in chronological order.

John Rowning, a fellow of Magdalene College, was born in 1701 and died in London in 1771. He wrote *A compendious system of natural philosophy*, published in two volumes in 1738; a treatise on the method of *fluxions*, published in 1756; and a description of a machine for solving equations, published in the *Philosophical transactions* for 1770.

Francis Wollaston, a fellow of Sidney College, who was born on Nov. 23, 1731, and took his B.A. as second wrangler in 1758, wrote several papers and works on practical astronomy; a list of these is given in Poggendorff's *Handwörterbuch*. He died at Chiselhurst on Oct. 31, 1815.

George Atwood was born in 1746, was educated at Westminster School, took his B.A. as third wrangler and first Smith's prizeman in 1769, and subsequently was elected a fellow and tutor of Trinity College. The inefficiency of the professorial body served as a foil to his lectures, which attracted all the mathematical talent of the university. They were not only accurate and clear, but delivered fluently and illustrated with great ingenuity. The apparatus for calculating the numerical value of the acceleration produced by gravity which is still known by his name was invented by him and used in his Trinity lectures in 1782 and 1783. Analyses of the courses delivered in 1776 and in 1784 were issued by him, and are still extant. Pitt attended Atwood's lectures, and was so much interested in them that he gave him a post in London; and for the last twenty years of his life Atwood was the financial adviser of every successive government. Atwood died in London on July 11, 1807.

His most important work was one on dynamics, published at Cambridge in 1784. He also wrote a treatise on the theory of arches published in 1804. Besides these he contributed several papers to the *Philosophical transactions*: these include one in 1781 on the theory of the sextant; one in 1794 on the mathematical theory of the watch, especially the times of vibration of balances; one in 1796, to which the Copley medal was awarded, on the positions of equilibrium of floating bodies; and lastly one in 1798 on the stability of ships.

Waring's rival for the Lucasian chair was **Francis Maseres**[1], a fellow of Clare Hall. Maseres was descended from a family of French Huguenots who had settled in England: he was born in London on Dec. 15, 1731, and took his B.A. as senior wrangler in 1752. After failing to be elected to the professorship he went to the bar, and subsequently as attorney-general to the province of Canada; on his return in 1773 he was made a cursitor baron of the Exchequer, and held that office till his death on May 19, 1824. In 1750 he published a trigonometry, and at a later time several tracts on algebra and the theory of equations: these are of no value, as he refused to allow the use of negative or impossible quantities. In 1783 he wrote a treatise in two volumes on the theory of life assurance, which is a creditable attempt to put the subject on a scientific basis. He has however acquired considerable celebrity from the reprints of most of the works either on logarithms or on optics by mathematicians of the seventeenth century, including those by Napier, Snell, Descartes, Schooten, Huygens, Barrow, and Halley. These were published in six volumes, 1791—1807, at his expense after a careful revision of the text under the titles *Scriptores logarithmici* and *Scriptores optici*.

Nevil Maskelyne was born in London on Oct. 6, 1732, was educated at Westminster School, and took his B.A. as seventh wrangler in 1754, and was subsequently elected to a fellowship at Trinity. In 1765 he succeeded Bliss at Greenwich as astronomer-royal: the rest of his life was given up to practical astronomy. The issue of the *Nautical almanack* was wholly due to him, and began in 1767; in 1772 he made the Schehallien observations from which he calculated (then for

[1] An account of his life is given in the *Gentleman's magazine* for June, 1824: see also pp. 121—3 of the *Budget of paradoxes* by A. De Morgan, London, 1872.

the first time) the mean density of the earth; lastly in 1790 he published the earliest standard catalogue of stars, and Delambre for that reason considers modern observational astronomy to date from that year. A list of his numerous papers contributed to the *Philosophical transactions* will be found in Poggendorff's *Handwörterbuch*. He died on Feb. 9, 1811.

Bewick Bridge, a fellow of Peterhouse and mathematical professor at Haileybury College, was born near Cambridge in 1767, graduated B.A. as senior wrangler in 1790, and died at Cherryhinton, of which he was vicar, on May 15, 1833. He wrote text-books on geometrical conics (two volumes, 1810), algebra (1810, 1815, and 1821), trigonometry (1810 and 1818), and mechanics (1813).

William Frend was born at Canterbury on Nov. 22, 1757, took his B.A. from Christ's College as second wrangler in 1780, and was subsequently elected to a fellowship in Jesus College. He published in 1796 a work entitled *Principles of algebra*, in which he rejected negative quantities as nonsensical. He is probably better known in connection with his banishment in 1793 from the university on account of his publication of a certain pamphlet called *Peace and Union*. I should add that he was only refused leave to reside, and was not deprived of his fellowship. Any sympathy for the harsh treatment which he seems to have experienced will probably be dissipated by reading his own account of the proceedings which he published at Cambridge in 1793. He died in London on Feb. 21, 1841.

John Brinkley, a fellow of Caius, and subsequently bishop of Cloyne, who was born in Suffolk in 1763 and graduated as senior wrangler and first Smith's prizeman in 1788, acquired considerable reputation as professor of astronomy at Dublin. He contributed numerous papers either to the Royal Society or to the corresponding society in Ireland on various problems in astronomy, also a few on different questions connected with the use of series. A complete list of these will be found in the *Catalogue of scientific papers from the year* 1800 issued by the Royal Society. He died in Dublin on Sept. 14, 1835.

Daniel Cresswell, a fellow of Trinity, who was born at Wakefield in 1776 and graduated as seventh wrangler in 1797, was a well-known "coach" of his day. In 1822 he took a college living, and died at Enfield on March 21, 1844. His most important works are the *Elements of linear perspective*, Cambridge, 1811; a translation of Venturoli's *Mechanics*, Cambridge, 1822; and a work on the geometrical treatment of problems of maxima and minima.

Miles Bland, a fellow and tutor of St John's College, who was born in 1786 and graduated as second wrangler in 1808, was one of the best known writers of elementary books at the beginning of the century: he went down from the university in 1823 and died in 1868. In 1812 he published a collection of algebraical problems, and in 1819 another of geometrical problems: these became well-known school books. In 1824 he issued an elementary work on hydrostatics; and this was followed in 1830 by a collection of mechanical problems.

James Wood, a fellow and subsequently the master of St John's College and dean of Ely, was born in Lancashire about 1760, graduated as senior wrangler in 1782, and died at Cambridge on April 23, 1839. His algebra was long a standard work, it formed originally a part of his *Principles of mathematics and natural philosophy* in four volumes, Cambridge, 1795—99; the section on astronomy (vol. iv. part ii.) was contributed by Vince. Wood also wrote a paper *On the roots of equations* which will be found in the *Philosophical transactions* for 1798.

It was with difficulty that I made out a list of some thirty or forty writers on mathematics of this time who were educated at Cambridge; and the above names comprise every one of them whose works can as far as I know be said to have influenced the development of the study at Cambridge or elsewhere.

It is not easy to make out exactly what books were usually read at this time, but Whewell says that they certainly included considerable parts of the *Principia*, the works of Cotes, Atwood, Vince, and Wood: the treatises by the two last-named mathematicians were probably read by all mathematical students.

Sir Frederick Pollock of Trinity, who was senior wrangler in 1806, in the account printed in the next paragraph, asserts that in his freshman's year he read Wood's *Algebra* (to quadratic equations), Bonnycastle's *Algebra*, and Simpson's *Euclid*: in his second year he read algebra beyond quadratic equations in Wood's work, and the theory of equations in the works by Wood and Vince: in his third year he read the Jesuit edition of Newton's *Principia*, Vince's *Fluxions*, and copied numerous manuscripts or analyses supplied by his coach. There is no doubt that he is right in saying that this was less than was usual.

The letter to which I have just referred was sent by Sir Frederick Pollock in July, 1869, to Prof. De Morgan in answer to a request for a trustworthy account, which would be of historical value, about the mathematical reading of men at the beginning of this century. It is so interesting that no excuse is necessary for reproducing it.

I shall write in answer to your inquiry *all* about my books, my studies, and my degree, and leave you to settle all about the proprieties which my letter may give rise to, as to egotism, modesty, &c. The only books I read the first year were Wood's *Algebra* (as far as quadratic equations), Bonnycastle's ditto, and *Euclid* (Simpson's). In the second year I read Wood (beyond quadratic equations), and Wood and Vince for what they called the *branches*. In the third year I read the *Jesuit's* Newton and Vince's *Fluxions*; these were all the *books*, but there were certain MSS. floating about which I copied—which belonged to Dealtry, second wrangler in Kempthorne's year. I have no doubt that I had read less and seen fewer books than any senior wrangler of about my time, or any period since; but what I knew I knew thoroughly, and it was completely at my fingers' ends. I consider that I was the last *geometrical* and *fluxional* senior wrangler; I was not up to the *differential* calculus, and never acquired it. I went up to college with a knowledge of Euclid and algebra to quadratic equations, nothing more; and I never read any second year's lore during my first year, nor any third year's lore during my second; my *forte* was, that what I *did* know I *could produce at any moment with* PERFECT *accuracy*. I could repeat the first book of Euclid word by word and letter by letter. During my first year I was not a '*reading*' man (so called); I had no expectation of honours or a fellowship, and I attended all the lectures on all subjects—Harwood's anatomical, Wollaston's chemical, and Farish's mechanical lectures—but the examination at the end of the first year revealed to me my powers. I was not only in the first class, but it was generally understood I was *first* in the first class; neither I nor any one for me expected I should get in at all. Now, as I had taken no pains to prepare (taking, however, marvellous pains while the examination was going on), I knew better than any one else the value of my *examination qualities* (great rapidity and perfect accuracy); and I said to myself, 'If you're not an ass, you'll be senior wrangler;' and *I took to 'reading' accordingly.* A curious circumstance occurred when the brackets[1] came out in the senate-house declaring the result of the examination: I saw at the top the name of Walter *bracketed alone* (as he was); in the bracket below were *Fiott, Hustler, Jephson.* I looked down and could not find my own name till I got to Bolland, when my pride took fire, and I said, 'I must have beaten *that man*, so I will look up again;' and on looking up carefully I found the nail had been passed through my name, and I was at the top bracketed *alone*, even above Walter. You may judge what my feelings were at this discovery; it is the only instance of two such brackets, and it made my fortune—that is, made me independent, and gave me an immense college reputation. It was said I was more than half of the examination before any one else. The two moderators were Hornbuckle, of St John's, and Brown (Saint Brown), of Trinity. The Johnian congratulated me. I said perhaps I might be challenged; he said, 'Well, if you are you're quite safe—you may sit down and do nothing, and no one would get up to you in a whole day.'.........

My experience has led me to doubt the value of competitive examination. I believe the most valuable qualities for practical life cannot be got at by any examination—such as steadiness and perseverance. It may be well to make an examination part of the mode of judging of a man's fitness; but to put him into an office with public duties to perform merely on his passing a good examination is, I think, a bad mode of preventing mere patronage. My brother is one of the best generals that

[1] The 'brackets' were a preliminary classification in order of merit. They were issued on the morning of the last day of the tripos examination. The names in each bracket were arranged in alphabetical order. A candidate who considered that he was placed too low in the list could challenge any one whose name appeared in the bracket next above that in which his own was placed, and if on re-examination he proved himself the equal of the man so challenged his name was transferred to the higher bracket (see p. 200).

ever commanded an army, but the qualities that make him so are quite beyond the reach of any examination. Latterly the Cambridge examinations seem to turn upon very different matters from what prevailed in my time. I think a Cambridge education has for its object to make good members of society—not to extend science and make profound mathematicians. The tripos questions in the senate-house ought not to go beyond certain limits, and geometry ought to be cultivated and encouraged much more than it is.

To this De Morgan replied:

Your letter suggests much, because it gives possibility of answer. The *branches* of algebra of course mainly refer to the second part of Wood, now called the theory of equations. Waring was his guide. Turner—whom you must remember as head of Pembroke, senior wrangler of 1767—told a young man in the hearing of my informant to be sure and attend to quadratic equations. 'It was a quadratic,' said he, 'made me senior wrangler.' It seems to me that the Cambridge *revivers* were Waring, Paley, Vince, Milner.

You had Dealtry's MSS. He afterwards published a very good book on fluxions. He merged his mathematical fame in that of a Claphamite Christian. It is something to know that the tutor's MS. was in vogue in 1800–1806.

Now—how did you get your conic sections? How much of Newton did you read? From Newton direct, or from tutor's manuscript?

Surely Fiott was our old friend Dr Lee. I missed being a pupil of Hustler by a few weeks. He retired just before I went up in February 1823. The echo of Hornbuckle's answer to you about the challenge has lighted on Whewell, who, it is said, wanted to challenge Jacob, and was answered that he could not beat [him] if he were to write the whole day and the other wrote nothing. I do not believe that Whewell would have listened to any such dissuasion.

I doubt your being the last fluxional senior wrangler. So far as I know, Gipps, Langdale, Alderson, Dicey, Neale, may contest this point with you.

The answer of Sir Frederick Pollock to these questions is dated August 7, 1869, and is as follows.

You have put together as *revivers* five very different men. Woodhouse was better than Waring, who could not prove Wilson's (Judge of C. P.) guess about the property of prime numbers; but Woodhouse (I think) did prove it, and a beautiful proof it is. Vince was a bungler, and I think utterly insensible of mathematical beauty.

Now for your questions. I did not get my conic sections from Vince. I copied a MS. of Dealtry's. I fell in love with the cone and its sections, and everything about it. I have never forsaken my favourite pursuit; I delighted in such problems as two spheres touching each other and also the inside of a hollow cone, &c. As to Newton, I read a good deal (men *now* read nothing), but I read much of the notes. I detected a blunder which nobody seemed to be aware of. Tavel, tutor of Trinity, was not; and he augured very favourably of me in consequence. The application of the Principia I got from MSS. The blunder was this: in calculating the resistance of a globe at the end of a cylinder oscillating in a resisting medium they had forgotten to notice that there is a difference between the resistance to a globe and a circle of the same diameter.

The story of Whewell and Jacob cannot be true. Whewell was a very, *very* considerable man, I think not a *great* man. I have no doubt Jacob beat him in accuracy, but the supposed answer *cannot* be true; it is a mere echo of what actually passed between me and Hornbuckle on the day the Tripos came out—for the truth of which I vouch. I think the examiners are taking too *practical* a turn; it is a waste of time to calculate *actually* a longitude by the help of logarithmic tables and lunar observations. It would be a fault not to know *how*, but a greater to be handy at it[1].

I may mention in passing that experimental physics began about this time to attract considerable attention. This was largely due to the influence of Cavendish, Young, W. H. Wollaston, Rumford, and Dalton in England, and of Lavoisier and Laplace in France. The first three of these writers came from Cambridge; and I add a few lines on the subject-matter of their works.

The honourable **Henry Cavendish**[2] was born at Nice on Oct. 10, 1731. His tastes for scientific research and mathematics seem to have been formed at Cambridge, where he resided from 1749 to 1753. He was a member of Peterhouse,

[1] *Memoir of A. De Morgan* (pp. 387—392), by S. E. De Morgan, London, 1882.

[2] An account of his life by G. Wilson will be found in the first volume of the publications of the Cavendish Society, London, 1851. His *Electrical researches* were edited by J. C. Maxwell, and published at Cambridge in 1879.

but like all fellow-commoners of the time did not present himself for the senate-house examination, and in fact he did not actually take a degree. He created experimental electricity, and was one of the earliest writers to treat chemistry as an exact science. In 1798 he determined the density of the earth by estimating its attraction as compared with that of two given lead balls: the result is that the mean density of the earth is about five and a half times that of water. This experiment was carried out in accordance with a suggestion which had been first made by John Michell, a fellow of Queens' [B.A. 1748], who had died before he was able to carry it into effect. His note-books prove him to have been much interested in mathematical questions but I believe he did not publish any of his results. He died in London on Feb. 24, 1810.

Thomas Young[1], born at Milverton on June 13, 1773, and died in London on May 10, 1829, was among the most eminent physicists of his time. He seems as a boy to have been somewhat of a prodigy, being well read in modern languages and literature as well as in science; he always kept up his literary tastes and it was he who first furnished the key to decipher the Egyptian hieroglyphics. He was destined to be a doctor, and after attending lectures at Edinburgh and Göttingen entered at Emmanuel College, Cambridge, from which he took his degree in 1803; and to his stay at the university he attributed much of his future distinction. His medical career was not particularly successful, and his favorite maxim that a medical diagnosis is only a balance of probabilities was not appreciated by his patients, who looked for certainty in return for their fee. Fortunately his private means were ample. Several papers contributed to various learned societies from 1798 onwards prove him to have been a mathematician of considerable power; but the researches which have immortalized his name are those by which he laid down the laws of interference of waves and of light, and was thus able to overcome

[1] For further details see his life and works by G. Peacock, 4 vols. 1855.

the chief difficulties in the way of the acceptance of the undulatory theory of light.

Another experimental physicist of the same time and school was **William Hyde Wollaston**, who was born at Dereham on Aug. 6, 1766, and died in London on Dec. 22, 1828. He was educated at Caius College (M.B. 1788), of which society he was a fellow. Besides his well-known chemical discoveries, he is celebrated for his researches on experimental optics, and for the improvements he effected in astronomical instruments.

One characteristic of this period to which I have not yet alluded is the rise of a class of teachers in the university who are generally known as coaches or private tutors, but I may conveniently defer any remarks on this subject until I consider the general question of the organization of education in the university (see pp. 160—163).

CHAPTER VII.

THE ANALYTICAL SCHOOL[1].

THE isolation of English mathematicians from their continental contemporaries is the distinctive feature of the history of the latter half of the eighteenth century. Towards the close of that century the more thoughtful members of the university recognized that this was a serious evil, and it would seem that the chief obstacle to the adoption of analytical methods and the notation of the differential calculus arose from the professorial body and the senior members of the senate, who regarded any attempt at innovation as a sin against the memory of Newton.

I propose in this chapter to give a sketch of the rise of the analytical school, and shall briefly mention the chief works of Robert Woodhouse, George Peacock, Charles Babbage, and Sir John Herschel. The later history of that school is too near our own times to render it possible or desirable to discuss it in similar detail: and I shall make no attempt to do so.

The earliest attempt in this country to explain and advocate the notation and methods of the calculus as used on the continent was due to Woodhouse, who stands out as the apostle of the new movement.

[1] For the few biographical notes given in this chapter I am generally indebted to the obituary notices which are printed in the transactions of the Royal and other similar learned societies.

Robert Woodhouse[1] was born at Norwich on April 28, 1773, took his B.A. as senior wrangler and first Smith's prizeman in 1795 from Caius College, was elected to a fellowship in due course, and continued to live at Cambridge till his death on Dec. 23, 1827.

His earliest work, entitled the *Principles of analytical calculation*, was published at Cambridge in 1803. In this he explained the differential notation and strongly pressed the employment of it, but he severely criticized the methods used by continental writers, and their constant assumption of non-evident principles. Woodhouse was a brilliant logician, but, perhaps partly for that reason, the style of the book is very crabbed; and it is difficult to read, on account of the extraordinary complications of grammatical construction in which he revels. This was followed in 1809 by a trigonometry (plane and spherical), and in 1810 by a historical treatise on the calculus of variations and isoperimetrical problems. He next produced an astronomy: the first volume (usually bound in two) on practical and descriptive astronomy being issued in 1812, the second volume, containing an account of the treatment of physical astronomy by Laplace and other continental writers, being issued in 1818. All these works deal critically with the scientific foundation of the subjects considered—a point which is not unfrequently neglected in modern text-books.

In 1820 Woodhouse succeeded Milner as Lucasian professor, but in 1822[2] he resigned it in exchange for the Plumian chair. The observatory at Cambridge was finished in 1824, and Woodhouse was appointed superintendent, but his health was then rapidly failing, though he lingered on till 1827.

[1] See the *Penny Cyclopaedia*, vol. XXVII.

[2] It will be convenient to state here that Woodhouse's successor in the Lucasian chair was *Thomas Turton*, of St Catharine's College. Turton was born in 1780 and graduated as senior wrangler in 1805. I am not aware that he ever lectured. In 1826 he exchanged the chair for one of divinity; in 1842 he was made dean of Westminster; and in 1845 bishop of Ely. He died in 1864.

A man like Woodhouse, of scrupulous honour, universally respected, a trained logician, and with a caustic wit, was well fitted to introduce a new system. "The character," says De Morgan, "which must be given of the several writings of Woodhouse entitles us to suppose that the revolution in our mathematical studies, of which he was the first promoter, would not have been brought about so easily if its earliest advocacy had fallen into less judicious hands. For instance, had he not, when he first called attention to the continental analysis, exposed the unsoundness of some of the usual methods of establishing it more like an opponent than a partizan, those who were averse from the change would probably have made a successful stand against the whole upon the ground which, as it was, Woodhouse had already made his own. From the nature of his subjects, his reputation can never equal that of the first seer of a comet with the world at large: but the few who can appreciate what he did will always regard him as one of the most philosophical thinkers and useful guides of his time."

Woodhouse's writings were of no use for the public examinations and were scouted by the professors, but apparently they were eagerly studied by a minority of students. Herschel[1], with perhaps a pardonable exaggeration, describes the general feeling of the younger members of the university thus. "Students at our universities, fettered by no prejudices, entangled by no habits and excited by the ardour and emulation of youth, had heard of the existence of masses of knowledge from which they were debarred by the mere accident of position. They required no more. The prestige which magnifies what is unknown, and the attractions inherent in what is forbidden, coincided in their impulse. The books were procured and read, and produced their natural effects. The brows of many a Cambridge moderator were elevated, half in ire, half in admiration, at the unusual answers which began to appear

[1] The reader will find another account by Whewell of the same movement in Todhunter's edition of his life (vol. II. pp. 16, 29, 30).

in examination papers. Even moderators are not made of impenetrable stuff: their souls were touched, though fenced with seven-fold Jacquier, and tough bull-hide of Vince and Wood."

But while giving Woodhouse all the credit due to his initiation, I doubt whether he exercised much influence on the majority of his contemporaries, and I think the movement might have died away for the time being, if the advocacy of Peacock had not given it permanence. I allude hereafter very briefly to him and others of those who worked with him. I will only say here that in 1812 three undergraduates—Peacock, Herschel, and Babbage—who were impressed by the force of Woodhouse's remarks and were in the habit of breakfasting together every Sunday morning, agreed to form an Analytical Society, with the object of advocating the general use in the university of analytical methods and of the differential notation, and thus as Herschel said "do their best to leave the world wiser than they found it." The other original members were William Henry Maule of Trinity, senior wrangler in 1810 and subsequently a justice of the common pleas, Thomas Robinson of Trinity, thirteenth wrangler in 1813, Edward Ryan of Trinity, who took his B.A. in 1814, and Alexander Charles Louis d'Arblay of Christ's, tenth wrangler in 1818. In 1816 the Society published a translation of Lacroix's *Elementary differential calculus*.

In 1817 Peacock, who was moderator for that year, introduced the symbols of differentiation into the papers set in the senate-house examination. But his colleague, John White of Caius (B.A. 1808), continued to use the fluxional notation. Peacock himself wrote on March 17 of 1817 (i.e. just after the examination) on the subject as follows: "I assure you that I shall never cease to exert myself to the utmost in the cause of reform, and that I will never decline any office which may increase my power to effect it. I am nearly certain of being nominated to the office of moderator in the year 1818–19, and as I am an examiner in virtue of my office, for the next year I shall pursue a course even more decided than hitherto, since I feel that men have been prepared for the change, and will then be enabled to have acquired a better system by the publication of improved elementary books. I have considerable influence as a lecturer, and I will not neglect it. It is by silent perseverance only that we can hope to reduce the many-headed monster of prejudice, and make the university answer her character as the loving mother of good learning and science."

The action of G. Peacock and the translation of Lacroix's treatise were severely criticised by D. M. Peacock in a work which was published at the expense of the university in 1819. The reformers were however encouraged by the support of most of the younger members of the university; and in 1819 G. Peacock, who was again moderator, induced his colleague Richard Gwatkin of St John's (B.A. 1814) to adopt the new notation. It was employed in the next year by Whewell[1], and in the following year by Peacock again, by which time the notation was well-established[2]: and subsequently the language of the fluxional calculus only appeared at rare intervals in the examination. It should however be noted in passing that it was only the exclusive use of the fluxional notation that was so hampering, and in fact the majority of modern writers use both systems. It was rather as the sign of their isolation and of the practice of treating all questions by geometry that the fluxional notation offended the reformers, than on account of any inherent defects of its own.

The Analytical Society followed up this rapid victory by

[1] Whewell gave but a wavering support to Peacock's action so long as its success was doubtful: see vol. II. p. 16, of Todhunter's *Life of Whewell*, London, 1876.

[2] A letter by Sir George Airy describing his recollections of the senate-house examination of 1823 and the introduction of analysis into the university examinations is printed in the number of *Nature* for Feb. 24, 1887. I think the contemporary statements of Herschel, Peacock, Whewell, and the criticisms of De Morgan, shew that the analytical movement was somewhat earlier than the time mentioned by Sir George Airy.

the issue in 1820 of two volumes of examples illustrative of the new method: one by Peacock on the differential and integral calculus, and the other by Herschel on the calculus of finite differences. Since then all elementary works on the subject have abandoned the exclusive use of the fluxional notation. But of course for a few years the old processes continued to be employed in college lecture-rooms and examination papers by some of the senior members of the university.

Amongst those who materially assisted in extending the use of the new analysis were Whewell and Airy. The former issued in 1819 a work on mechanics, and the latter, who was a pupil of Peacock, published in 1826 his *Tracts*, in which the new method was applied with great success to various physical problems. Finally, the efforts of the society were supplemented by the publication by Parr Hamilton in 1826 of an analytical geometry, which was an improvement on anything then accessible to English readers.

The new notation had barely been established when a most ill-advised attempt[1] was made to introduce another system, in which $\frac{dy}{dx}$ was denoted by $d_x y$. This was for some years adopted in the Johnian lecture-rooms and examination papers, but fortunately the strong opposition of Peacock and De Morgan prevented its further spread in the university. In fact uniformity of notation is essential to freedom of communication, and one would have supposed that those who admitted the evil of the isolation to which Cambridge and England had for a century been condemned would have known better than to at once attempt to construct a fresh language for the whole mathematical world.

[1] See *On the notation of the differential calculus*, Cambridge, 1832: and also the article by A. De Morgan in the *Quarterly journal of education* for 1834. De Morgan says it was first used in Trinity, but I can find no trace of it in the examination papers of that college. It occurs in the papers set in the annual examination at St John's in the years 1830, 1831, and 1832. I suspect that it was invented by Whewell, but I have no definite evidence of the fact.

The use of analytical methods spread from Cambridge over the rest of the country, and by 1830 they had almost entirely superseded the fluxional and geometrical methods. It is possible that the complete success of the new school and the brilliant results that followed from their teaching led at first to a somewhat too exclusive employment of analysis; and there has of late been a tendency to revert to graphical and geometrical processes. That these are useful as auxiliaries to analysis, that they afford elegant demonstrations of results which are already known, and that they enable one to grasp the connection between different parts of the same subject is universally admitted, but it has yet to be proved that they are equally potent as instruments of research. To that I may add, that in my opinion the analytical methods are peculiarly suited to the national genius.

I have often thought that an interesting essay might be written on the influence of race in the selection of mathematical methods. The Semitic races had a special genius for arithmetic and algebra, but as far as I know have never produced a single geometrician of any eminence. The Greeks on the other hand adopted a geometrical procedure wherever it was possible, and they even treated arithmetic as a branch of geometry by means of the device of representing numbers by lines. In the modern and mixed races of Europe the effects are more complex, but I think until Newton's time English mathematics might be characterized as analytical. Some admirable text-books on arithmetic and algebra were produced, and the only three writers previous to Newton who shewed marked original power in pure mathematics—Briggs, Harriot, and Wallis—generally attacked geometrical problems by the aid of algebra or analysis. For more than a century the tide then ran the other way; and the methods of classical geometry were everywhere used. This was wholly due to Newton's influence, and as with the lapse of time that died away the analytical methods again came into favour.

I add a few notes on the writers above-mentioned and their immediate successors, but with the establishment of the analytical school I consider my task is finished.

George Peacock, who was the most influential of the early members of the new school, was born at Denton on April 9, 1791, and took his B.A. from Trinity as second wrangler and second Smith's prizeman in 1813. He was elected to a fellowship in 1814, and subsequently was made a tutor of the college. I have already alluded to the prominent part which he took in introducing analysis into the senate-house examination.

Of his work as a tutor there seems to be but one opinion. An old pupil, himself a man of great eminence, says, "While his extensive knowledge and perspicuity as a lecturer maintained the high reputation of his college, and commanded the attention and admiration of his pupils, he succeeded to an extraordinary degree in winning their personal attachment by the uniform kindliness of his temper and disposition, the practical good sense of his advice and admonitions, and the absence of all moroseness, austerity, or needless interference with their conduct." "His inspection of his pupils," says another of them, "was not minute, far less vexatious; but it was always effectual, and at all critical points of their career, keen and searching. His insight into character was remarkable."

The establishment of the university observatory was mainly due to his efforts. In 1836 he was appointed to the Lowndean professorship in succession to W. Lax (see p. 105). The rival candidate was Whewell. In 1839 Peacock was made dean of Ely, and resided there till his death on Nov. 8, 1858.

Although Peacock's influence on the mathematicians of his time and his pupils was very considerable he has left few remains. The chief are his *Examples illustrative of the use of the differential calculus*, 1820; his article on *Arithmetic* in the *Encyclopaedia Metropolitana*, 1825, which contains the best historical account of the subject yet written, though the arrangement is bad; his *Algebra*, 1830 and 1842; and his *Report on recent progress in analysis*, 1833, which commenced those valuable summaries of scientific progress which enrich many of the annual volumes of the British Association.

The next most important member of the Analytical Society was **Charles Babbage**[1], who was born at Totnes on Dec. 26, 1792, and died in London on Oct. 18, 1871. He entered at Trinity College in April, 1810, as a bye-term student and was thus practically in the same year as Herschel and Peacock. Before coming into residence Babbage was already a fair mathematician, having mastered the works on fluxions by Humphry Ditton, Maclaurin, and Simpson, Agnesi's *Analysis* (in the English translation of which by the way the fluxional notation is used), Woodhouse's *Principles of analytical calculation*, and Lagrange's *Théorie des fonctions*.

It was he who gave the name to the Analytical Society, which he stated was formed to advocate "the principles of pure *d*-ism as opposed to the *dot*-age of the university." The society published a volume of memoirs, Cambridge, 1813; the preface and the first paper (on continued products) are due to Babbage: this work is now very scarce.

Finding that he was certain to be beaten in the tripos by Herschel and Peacock, Babbage migrated in 1813 to Peterhouse and entered for a poll degree, in order that he might be first both in his college and his examination in the senate-house. After taking his B.A. he moved to London, and an inspection of the catalogue of scientific papers issued by the Royal Society shews how active and many-sided he was. The most important of his contributions to the *Philosophical transactions* seem to be those on the calculus of functions, 1815 to 1817, and the magnetisation of rotating plates, 1825. In 1823 he edited the *Scriptores optici* for baron Maseres (see p. 108). In 1820 the Astronomical Society was founded mainly through his efforts, and at a later time, 1830 to 1832, he took a prominent part in the foundation of the British Association.

In 1828 he succeeded Airy as Lucasian professor and held

[1] He left an autobiography under the title *Passages from the life of a philosopher*. London, 1864.

the chair till 1839, but by an abuse which was then possible he neither resided nor taught.

Babbage will always be famous for his invention of an analytical machine, which could not only perform the ordinary processes of arithmetic, but could tabulate the values of any function and print the results. The machine was never finished, but the drawings of it, now deposited at Kensington, satisfied a scientific commission that it could be constructed.

The third of those who helped to establish the new method was Herschel. **Sir John Frederick William Herschel** was born at Slough on March 7, 1792. His father was Sir William Herschel (1738—1822) who was the most illustrious astronomer of the last half of the last century. Two anecdotes of his boyish years were frequently told by him as illustrative of his home training, and are sufficiently interesting to deserve repetition. One day when playing in the garden he asked his father what was the oldest thing with which he was acquainted. His father replied in Socratic manner by asking what the lad thought "was the oldest of all things." The replies were all open to objection, and finally the astronomer answered the question by picking up a stone and saying that that was the oldest thing of which he had definite knowledge. On another occasion in a conversation he asked the boy what sort of things were most alike. After thinking it over young Herschel replied that the leaves of a tree were most like one another. "Gather then a handful of leaves from that tree," said the philosopher, "and choose two that are alike." Of course it was impossible to do so. Both stories are trivial, but they were typical of the manner in which he was brought up, and these two particular incidents happened to make a deep impression on his mind.

Except for one year spent at Eton he was educated at home. In 1809 he entered at St John's College, graduating as senior wrangler and first Smith's prizeman in 1813.

His earliest original work was a paper on Cotes's theorem, which he sent when yet an undergraduate to the Royal Society, and immediately after taking his degree it was followed by others on mathematical analysis. He went down from the university in or about 1816, and for a few years read for the bar; but his natural bent was to chemistry and astronomy, and to those he soon turned his exclusive attention. The desire to complete his father's work led ultimately to his taking up the latter rather than the former subject. He died at Collingwood on May 11, 1871.

Besides his numerous papers on astronomy, his *Outlines of astronomy* published in 1849, and his articles on *Light* and *Sound* in the *Encyclopaedia Metropolitana* appear to be the most important of his contributions to science. His addresses to the Astronomical and other societies have been republished, and throw considerable light on the problems of his time. His *Lectures on familiar subjects* published in 1868 are models of how the mathematical solutions of physical and astronomical problems can be presented in an accurate manner and yet be made intelligible to all readers.

Another member of the university who took a prominent part in developing the study of analytical methods was Whewell. **William Whewell**[1], of Trinity College, was born at Lancaster on May 24, 1794, graduated as second wrangler and second Smith's prizeman in 1816, and was in due course elected to a fellowship. His life was spent in the work of his college and university. He was tutor of Trinity from 1823 to 1839, and master from 1841 to his death in 1866; while at different times he held in the university the chairs of mineralogy and moral philosophy.

His chief original works were his *History of the inductive sciences* and his papers on the tides, for the latter of which he received a medal of the Royal Society; but for my purpose he is chiefly noticeable for the great influence he exerted on his contemporaries.

[1] Two accounts of his life have been written: one by I. Todhunter in two volumes, London, 1876; and the other by Stair Douglas, London, 1881. The more important facts form the subject of an appreciative and graceful article by W. G. Clark in *Macmillan's magazine* for April, 1866.

Whewell occupied to his generation somewhat the same position that Bentley had done to the Cambridge of his day. But though Whewell was almost as masterful and combative as Bentley he was honest, generous, and straightforward. He lived to see his unpopularity pass away, his wonderful attainments universally recognized, and to enjoy the hearty respect of all and the love of many. His contemporaries seem to have regarded him as the most striking figure of the present century, but his range of knowledge was so wide and discursive that it could not be very deep, and his reputation has faded with great rapidity. Perhaps a future generation will rate him more highly than that of to-day, though he will always occupy a prominent position in the history of the university and his college.

With a view of stimulating still further the interest in mathematical and scientific subjects and the new methods of treating them, a permanent association known as the Cambridge Philosophical Society was established in 1819. It proved very useful, and noticeably so during the first twenty or thirty years after its formation. It was incorporated in 1832.

The character of the instruction in mathematics at the university has at all times largely depended on the text-books then in use. The importance of good books of this class has been emphasized by a traditional rule that questions should not be set on a new subject in the tripos unless it had been discussed in some treatise suitable and available for Cambridge students. Hence the importance attached to the publication of the work on analytical trigonometry by Woodhouse in 1809, and of the works on the differential calculus by the Analytical Society in 1816 and 1820. It will therefore be advisable to enumerate here some of the mathematical text-books brought out by members of the new school. I generally confine myself to those published before 1840, and thus exclude the majority of those known to undergraduates of the present day.

Wallis had published a treatise on analytical conic sections in 1665, but it had fallen out of use; and the only work on the subject commonly read at Cambridge at the beginning of the century was an appendix of about thirty pages at the end of Wood's *Algebra*. This was headed *On the application of algebra to geometry*, and it contained the equations of the straight line, ellipse, and a few other curves, rules for the construction of equations, and similar problems.

The senate-house papers from 1800 to 1820 shew that at the beginning of the century analytical geometry was always represented to some extent, though scarcely as an independent subject. Most of the questions relate to areas and loci, in which little more than the mode of representation by means of abscissæ and ordinates are involved. Even as late as 1830 the editor of the ninth edition of Wood's *Algebra* deemed that the chapter above mentioned afforded a sufficient account of the subject.

The need of a text-book on analytical geometry was first supplied by the work by Henry Parr Hamilton issued in 1826, and above alluded to. Hamilton was born at Edinburgh on April 3, 1794, and graduated from Trinity College as ninth wrangler in 1816; he was elected in due course to a fellowship, and held various college offices. He went down in 1830. In 1850 he was appointed dean of Salisbury, and lived there till his death on Feb. 7, 1880. In 1826 Hamilton published his *Principles of analytical geometry*, in which he defined the conic sections by means of the general equation of the second degree, and discussed the elements of solid geometry. Two years later, in 1828, he supplemented this by another and more elementary work, termed *An analytical system of conic sections*, in which he defined the curves by the focus and directrix property, as had been first suggested by Boscovich: the latter of these books went through numerous editions, and was translated into German.

In 1830 John Hymers (of St John's, second wrangler in 1826, died in 1887) published his *Analytical geometry of three dimensions.* In 1833 Peacock issued (anonymously) a *Syllabus of trigonometry, and the application of algebra to geometry*, seventy pages of which are devoted to analytical geometry; there was a second edition in 1836. Hymers's *Conic sections* appeared in 1837; it superseded Hamilton's in the university, and remained the standard work until the publication of the text-books still in use.

Among works on the calculus subsequent to those of Peacock and Herschel I should mention one by Thomas Grainger Hall (of Magdalene College, fifth wrangler in 1824, and subsequently professor of mathematics at King's College, London), issued in 1834, and the work by De Morgan published in 1842. Henry Kuhff, of St Catharine's (B.A. 1830, died in 1842), issued a work on finite differences in 1831; but I have never seen a copy of it. In 1841 a *Collection of examples illustrative of the use of the calculus* was published by Duncan Farquharson Gregory, a fellow of Trinity College: this was a work of great ability and was one of the earliest attempts to bring the calculus of operations into common use. Gregory was born at Edinburgh in April, 1813, graduated as fifth wrangler in 1837, and died on Feb. 23, 1844. His writings, edited by W. Walton, accompanied by a biographical memoir by R. L. Ellis[1], were published at Cambridge in 1865.

There was not the same need in applied mathematics for a new series of text-books, since optics, hydrostatics, and astronomy were already fairly represented, and Woodhouse's work on the latter involved the analytical discussion of dynamics. There was however no good work on elementary mechanics, and one was urgently required: this was supplied by the publication in 1819 of Whewell's *Mechanics*, and in 1823 of the same author's *Dynamics.* Another text-book on the subject was the translation of Venturoli's *Mechanics* by D. Cresswell,

[1] *Robert Leslie Ellis*, of Trinity College, who was born at Bath in 1817 and died at Cambridge in 1859, was senior wrangler in 1840. His memoirs were collected and published in 1863, and a life by H. Goodwin, the present bishop of Carlisle, is prefixed to them.

issued in 1822 (see p. 110). In 1832–34 Whewell re-issued his *Dynamics* in a greatly enlarged form and in three parts, and in 1837 published the *Mechanical Euclid.* Most of the older text-books in hydrostatics were superseded by Bland's *Elements of hydrostatics*, published in 1824.

In 1823 Henry Coddington, of Trinity College (who was senior wrangler in 1820 and died at Rome on March 3, 1845), issued a text-book on geometrical optics, which was practically a transcript of Whewell's lectures in Trinity on the subject. In 1838 William Nathaniel Griffin (senior wrangler in 1837) published his *Optics*, and this remained for many years a standard work. In 1829 Coddington issued a treatise on physical optics, which was followed by papers on various problems in that subject.

The publication by Sir George Airy of his *Tracts* in 1826 exercised a far greater influence on the study of mathematical physics in the university than the works just mentioned. A second edition of the *Tracts*, which appeared in 1831, contained a chapter on the *Undulatory theory of light*, a subject which was thenceforth freely represented in the tripos.

I should add to the above remarks that between 1823 and 1830 Dionysius Lardner (born in 1793 and died in 1859) brought out a series of treatises on the greater number of the subjects above mentioned.

From 1840 onwards an immense number of text-books were issued. I do not propose to enumerate them, but I may in passing just allude to the works on most of the subjects of elementary mathematics brought out at a somewhat later date by Isaac Todhunter, of St John's College, who was born at Rye in 1820, graduated as senior wrangler in 1848, and died at Cambridge in 1884. His text-books, if somewhat long, were always reliable, and for some years they were in general use. Besides these Todhunter wrote histories of the calculus of variations, of the theory of probabilities, and of the theory of attractions.

It would be an invidious task to select a few out of the

roll of eminent mathematicians who have been educated at Cambridge under the analytical school. But the names of those who have held important mathematical chairs will serve to shew how powerful that school has been, and confining myself strictly to the above, and omitting any reference to others—no matter how influential—I may just mention the following names as a sort of appendix to this chapter. The order in which they are arranged is determined by the dates of the tripos lists. I add a few remarks on the works of Augustus De Morgan, George Green, and James Clerk Maxwell, but in general I confine myself to giving the name of the professor and mentioning the chair that he held or holds.

The senior wrangler in the tripos of 1819 was **Joshua King**, of Queens' College, who was born in 1798 and died in 1857. King was Lucasian professor from 1839 to 1849 in succession to Babbage.

Sir George Biddell Airy, of Trinity College, who was senior wrangler in 1823, was born in Northumberland on July 27, 1801. In 1826 he succeeded Thomas Turton in the Lucasian chair, which in 1828 he exchanged for the Plumian professorship, where he followed Woodhouse: he held this professorship until his appointment as astronomer-royal in 1836, in succession to John Pond.

The senior wrangler of 1825 was **James Challis**, of Trinity, who was born in 1803 and died on Dec. 3, 1882: Challis was Plumian professor in succession to Sir George Airy from 1836 to 1882.

The year 1827 is marked by the name of **Augustus De Morgan**[1], who graduated from Trinity as fourth wrangler. He was born in Madura (Madras) in June 1806. In the then state of the law he was (as a unitarian) unable to stand for a fellowship, and accordingly in 1828 he accepted the chair of mathematics at the newly-established university of London, which is the same institution as that now known as Uni-

[1] His life has been written by his widow S. E. De Morgan. London, 1882.

versity College. There (except for five years from 1831 to 1835) he taught continuously till 1867, and through his works and pupils exercised a wide influence on English mathematics. The London Mathematical Society was largely his creation, and he took a prominent part in the proceedings of the Royal Astronomical Society. He died in London on March 18, 1871.

He was perhaps more deeply read in the philosophy and history of mathematics than any of his contemporaries, but the results are given in scattered articles which well deserve collection and republication. A list of these is given in his life, and I have made considerable use of some of them in this book. The best known of his works are the memoirs on the foundation of algebra, *Cambridge philosophical transactions*, vols. VIII. and IX.; his great treatise on the differential calculus published in 1842, which is a work of the highest ability; and his articles on the calculus of functions and on the theory of probabilities in the *Encyclopaedia Metropolitana.* The article on the calculus of functions contains an investigation of the principles of symbolic reasoning, but the applications deal with the solution of functional equations rather than with the general theory of functions. The article on probabilities gives a very clear analysis of the mathematics of the subject to the time at which it was written.

In 1830 we have the names of **Charles Thomas Whitley**, subsequently professor of mathematics at the university of Durham; **James William Lucas Heaviside**, subsequently professor of mathematics at the East India College, Haileybury; and **Charles Pritchard**, now Savilian professor of astronomy at the university of Oxford.

In 1837 the second wrangler was **James Joseph Sylvester**, who is now Savilian professor of geometry at the university of Oxford. Among the numerous memoirs he has contributed to learned societies I may in particular single out those on canonical forms, the theory of contravariants, reciprocants, the theory of equations, and lastly that on Newton's rule. He has also created the language and notation of considerable parts of the various subjects on which he has written.

In the same list appears the name of **George Green**, who was one of the most remarkable geniuses of this century. Green was born near Nottingham in 1793. Although self-educated he contrived to obtain copies of the chief mathematical works of his time. In a paper of his, written in 1827 and published by subscription in the following year, the term potential was first introduced, its leading properties proved, and the results applied to magnetism and electricity. In 1832 and 1833 papers on the equilibrium of fluids and on attractions in space of n dimensions were presented to the Cambridge Philosophical Society, and in the latter year one on the motion of a fluid agitated by the vibrations of a solid ellipsoid was read before the Royal Society of Edinburgh. In 1833 he entered at Caius College, graduated as fourth wrangler in 1837, and in 1839 was elected to a fellowship. Directly after taking his degree he threw himself into original work, and produced in 1837 his paper on the motion of waves in a canal, and on the reflexion and refraction of sound and light. In the latter the geometrical laws of sound and light are deduced by the principle of energy from the undulatory hypothesis, the phenomenon of total reflexion is explained physically, and certain properties of the vibrating medium are deduced. In 1839, he read a paper on the propagation of light in any crystalline medium. All the papers last named are printed in the *Cambridge philosophical transactions* for 1839. He died at Cambridge in 1841. A collected edition of his works was published in 1871.

The senior wrangler in 1841 was **George Gabriel Stokes**, of Pembroke College, who was born in Sligo on Aug. 13, 1819, and in 1849 succeeded Joshua King as Lucasian professor. In the following year **Arthur Cayley**, of Trinity College, was senior wrangler: he was born at Richmond, Surrey, on Aug. 16, 1821, and in 1863 was appointed Sadlerian professor. In the tripos of the next year **John Couch Adams**, of St

John's College, and now of Pembroke College, was senior wrangler: he was born in Cornwall on June 5, 1819, and in 1858 succeeded Peacock as Lowndean professor.

The second wrangler in 1843 was **Francis Bashforth**, who was subsequently appointed professor at Woolwich. His researches, especially those on the motion of a projectile in a resisting medium (London, 1873), have been and are in constant use among artillerymen and engineers of all nations.

The second wrangler in 1845 was **Sir William Thomson**, of Peterhouse, who was born at Belfast in June, 1824, and is now professor of natural philosophy at the university of Glasgow. I need hardly say here that Sir William Thomson has enriched every department of mathematical physics by his writings. His collected papers are now being published by the university of Cambridge. Among other names in the same tripos are those of **Hugh Blackburn**, of Trinity College, who was subsequently professor of mathematics at the university of Glasgow, and of **George Robarts Smalley**, the astronomer-royal of New South Wales.

The senior wrangler of 1852 was **Peter Guthrie Tait**, now professor of natural philosophy at the university of Edinburgh, who besides other well-known works was joint author with Sir William Thomson of the epoch-marking *Treatise on natural philosophy*, of which the first edition was published in 1867.

The year 1854 is distinguished by the name of **James Clerk Maxwell**, of Trinity College, who was second wrangler; **Edward James Routh**, of Peterhouse, being senior wrangler. Maxwell[1] was born in Edinburgh on June 13, 1831. His earliest paper was written when only fourteen on a mechanical method of tracing cartesian ovals, and was sent to the Royal Society of

[1] A tolerably full account of his life and a list of his writings will be found either in vol. XXIII. of the *Proceedings* of the Royal Society, or in the article contributed by Prof. Tait to the *Encyclopaedia Britannica*. For fuller details, his life by L. Campbell and W. Garnett, London, 1882, may be consulted. His collected works are being edited by Prof. Niven, and will shortly be published by the university of Cambridge.

Edinburgh. His next paper written three years later was on the theory of rolling curves, and was immediately followed by another on the equilibrium of elastic solids. At Cambridge in 1854 after taking his degree he read papers on the transformation of surfaces by bending, and on Faraday's lines of force. These were followed in 1859 by the essay on the stability of Saturn's rings, and various articles on colour. He held a chair of mathematics at Aberdeen from 1856 to 1860; and at King's College, London, from 1860 to 1868; in 1871 he was appointed to the Cavendish chair of physics at Cambridge. His most important subsequent works were his *Electricity and magnetism* issued in 1873, his *Theory of heat* published in 1871, and his elementary text-book on *Matter and motion.* To these works I may add his papers on the molecular theory of gases and the articles on cognate subjects which he contributed to the ninth edition of the *Encyclopaedia Britannica.* He died at Cambridge on Nov. 5, 1879.

His *Electricity and magnetism*, in which the results of various papers are embodied, has revolutionized the treatment of the subject. Poisson and Gauss had shewn how electrostatics might be treated as the effects of attractions and repulsions between imponderable particles; while Sir William Thomson in 1846 had shewn that the effects might also and with more probability be supposed analogous to a flow of heat from various sources of electricity properly distributed. In electro-dynamics the only hypothesis then current was the exceedingly complicated one proposed by Weber, in which the attraction between electric particles depends on their relative motion and position. Maxwell rejected all these hypotheses and proposed to regard all electric and magnetic phenomena as stresses and motions of a material medium; and these, by the aid of generalized coordinates, he was able to express in mathematical language. He concluded by shewing that if the medium were the same as the so-called luminiferous ether, the velocity of light would be equal to the ratio of the electromagnetic and electrostatic units. This appears to be the case, though these units have not yet been determined with sufficient precision to enable us to speak definitely on the subject.

Hardly less eventful, though less complete, was his work on the kinetic theory of gases. The theory had been established by the labours of Joule in England and Clausius in Germany; but Maxwell reduced it to a branch of mathematics. He was engaged on this subject at the time of his death, and his two last papers were on it. It has been the subject of some recent papers by Boltzmann.

In the tripos list of 1859 appear the names of **William Jack**, professor of mathematics at the university of Glasgow; of **Edward James Stone**, the Radcliffe observer at the university of Oxford; and of **Robert Bellamy Clifton**, the professor of physics at the university of Oxford.

I repeat again that the above list is in no way intended to be exhaustive, but is rather to be taken as one illustration of the growing numbers and reputation of the Cambridge school of mathematics.

The year at which I stop is the first of the Victorian statutes; and is a well-defined date at which I may close this history.

We live in an age somewhat analogous to that of the commencement of the renaissance. The system of education under the Elizabethan statutes—narrow in its range of studies and based on theological tests—has given way to one where subjects of all kinds are eagerly studied. The rise of the analytical school in mathematics and the establishment of the classical tripos in 1824 are the first outward and visible signs of the new intellectual activity which was quickening the whole life of the university. The mathematicians have taken their full share in that life, and that they have again raised Cambridge to the position of one of the chief mathematical schools of Europe will I think be admitted by the historian of the subsequent history of mathematics in Cambridge.

CHAPTER VIII.

THE ORGANIZATION AND SUBJECTS OF EDUCATION[1].

SECTION 1. *The mediaeval system of education.*
SECTION 2. *The period of transition.*
SECTION 3. *The system of education under the Elizabethan statutes.*

IN the preceding chapters I have enumerated most of the eminent mathematicians educated at Cambridge, and have indicated the lines on which the study of mathematics developed. I propose now to consider very briefly the kind of instruction provided by the university, and the means adopted for testing the proficiency of students.

Until 1858 the chief statutable exercises for a degree were the public maintenance of a thesis or proposition in the schools

[1] In writing this chapter I have mainly relied on *Observations on the statutes of the university of Cambridge* by G. Peacock, London, 1841, and on the *University of Cambridge* by J. Bass Mullinger, 2 volumes, Cambridge, 1873 and 1884. The most complete collection of documents referring to Cambridge is that contained in the *Annals of Cambridge* by C. H. Cooper, 5 volumes, Cambridge, 1842—52; but the collection of *Documents relating to the university and colleges of Cambridge*, issued by the Royal Commissioners in 1852, is for many purposes more useful. The *Statuta antiqua* are printed at the beginning of the edition of the statutes issued at Cambridge in 1785, and are reprinted in the *Documents*. It would seem from the *Munimenta academica* by Henry Anstey in the Rolls Series, London, 1848, that the customs at Oxford only differed in small details from those at Cambridge, and the regulations of either university may be used to illustrate contemporary student life at the other: but migration between them was so common that it would have been strange if it had been otherwise.

against certain opponents, and the opposition of a proposition laid down by some other student. Every candidate for a degree had to take part in a certain number of these discussions.

The subject-matter of these "acts" varied at different times. In the course of the eighteenth century it became the custom at Cambridge to "keep" some or all of them on mathematical questions, and I had at first intended to confine myself to reproducing one of the disputations kept in that century. But as the whole mediæval system of education—teaching and examining—rested on the performance of similar exercises, and as our existing system is derived from that without any break of continuity, I thought it might be interesting to some of my readers if I gave in this chapter a sketch of the course of studies, the means of instruction, and the tests imposed on students in earlier times; leaving the special details of a mathematical act to another chapter. It will therefore be understood that I am here only indirectly concerned with the history of the development of mathematical studies.

I also defer to a subsequent chapter the description of the origin and history of the mathematical tripos. I will only here remark that the university was not obliged to grant a degree to any one who performed the statutable exercises, and after the middle of the eighteenth century the university in general refused to pass a supplicat for the B.A. degree unless the candidate had also presented himself for the senate-house examination. That examination had its origin somewhere about 1725 or 1730, and though not recognized in the statutes or constitution of the university it gradually superseded the discussions as the actual test of the ability of students.

The mediaeval system of education.

The rules of some of the early colleges, especially those of Michael-house (founded in 1324, which now forms part of Trinity College), regulated every detail of the daily life of their members, and together with the ancient statutes of the university enable us to picture the ordinary routine of the career of a mediæval student.

In the thirteenth or fourteenth century then a boy came up to the university at some age between ten and thirteen under the care of a "fetcher," whose business it was to collect from some district about twenty or thirty lads and bring them up in one party. These "bringers of scholars" were protected by special enactments[1]. On his arrival the boy was generally entered under some master of arts who kept a hostel (i.e. a private boarding-house licensed by the university) or if very lucky got a scholarship at a college. The university in its corporate capacity did not concern itself much about the discipline or instruction of its younger members: times were rough and life was hard, and if one student more or less died or otherwise came to grief no one cared about it, so that a student who relied on the university alone or got into a bad hostel was in sorry straits.

If we follow the course of a student who was at one of the colleges or better hostels we may say that in general he spent the first four years of his residence in studying the subjects of the trivium, that is, Latin grammar, logic, and rhetoric. During that time he was to all intents a schoolboy, and was treated exactly like one. It is noticeable that the technical term for a student on presentation for the bachelor's degree is still *juvenis*, and the word *vir* is reserved for those who are at least full bachelors.

Few of those who thus came up knew anything beyond the merest elements of Latin, and the first thing a student had to learn was to speak, read, and write that language. It is probable that to the end of the fourteenth century the bulk of those who came to the university did not progress beyond this, and were merely students in grammar attending the glomerel schools. There would seem to have been nearly a dozen such

[1] *Munimenta academica*, 346; Lyte, 198.

schools in the thirteenth century, each under one master, and all under the supervision of a member of the university, known as the *magister glomeriae*[1]. This master of glomery had as such no special right over the other students of the university[2], but the "glomerels" were of course subject to his authority; and to enhance his dignity he had a bedell to attend him. To these glomerels the university gave the degree of "master in grammar," which served as a license to teach Latin, gave the coveted prefix of *dominus* or *magister* (which in common language was generally rendered *dan*, *don*, or *sir*), and distinguished the clerk from a mere "hedge-priest." To get this degree the glomerel had first to shew that he had studied Priscian in the original, and then to give a practical demonstration of proficiency in the mechanical part of his art. The regulations were that on the glomerel proceeding to his degree "then shall the bedell purvay for every master in grammar a shrewd boy, whom the master in grammar shall beat openly in the grammar schools, and the master in grammar shall give the boy a groat for his labour, and another groat to him that provideth the rod and the palmer, etcetera, *de singulis*. And thus endeth the act in that faculty[3]." The university presented the new master in grammar with a palmer, that is a ferule; he took a solemn oath that he would never teach Latin out of any indecent book; and he was then free of the exercise of his profession. The last degree in grammar was given in 1542. A student in grammar in general went down as soon as he got his degree. The resident masters in grammar occupied a very subordinate position in the university hierarchy. They not only yielded precedence to bachelors, but there were express

[1] Mullinger, I. 340.

[2] These rules were laid down in 1275 by Hugh Balsham, the bishop of Ely.

[3] The account of this and other ceremonies of the mediæval university is taken from the bedell's book compiled in the sixteenth century by Matthew Stokes, a fellow of King's and registrary of the university. It is printed at length in an appendix to Peacock's *Observations*.

statutes[1] that the university should not attend the funeral of one of them.

The corresponding degree of master of rhetoric was occasionally given. The last degree in this faculty was conferred in 1493.

Ambitious students or the scholars of a college were expected to know something of Latin before they came up; but the knowledge was generally of a very elementary character, and not more than could be picked up at a monastic or cathedral school. These lads formed the honour students, and took their degrees in arts.

To obtain the degree of master of arts in the thirteenth century it was necessary first to obtain a *licentia docendi*, and secondly to be "incepted," that is, admitted by the whole body of teachers or regents as one of themselves. The *licentia docendi* was originally obtained on proof of good moral character from the chancellor of the chapter of the church with which the university was in close connection. For inception the student was then recommended by a master of the university under whom he had studied, and the student had to keep an act or give a lecture before the whole university. On his inception he gave a dinner or presents to his new colleagues. Possibly the procedure was as elaborate as that described immediately hereafter, but we do not know any details beyond the above.

At a later time, as education became more general, the lads were somewhat older when they came up, and were already acquainted[2] with Latin grammar. The students in grammar thus gradually declined in numbers, and finally were hardly regarded as being members of the university. By the fifteenth century the average age at entrance was thirteen or four-

[1] *Statuta antiqua*, 178; *Documents*, I. 404. Similar regulations existed at Oxford, *Munimenta academica*, 264, 443.

[2] In founding King's College Henry VI. seems to have assumed that the scholars would have already mastered all the subjects of the trivium at Eton. The statute is quoted in Mullinger, I. 308.

teen[1], and most of the students proceeded in arts. From this time forward the *statuta antiqua* of the university enable us to sketch the course of a student in far greater detail, but there is no reason to suppose that it was substantially different from that of a student in arts in the two preceding centuries.

A student in arts spent the first year of his course in learning Latin. This at first meant Priscian and grammar only, but in the fifteenth century Terence, Virgil, and Ovid were added as text-books which should be used, and versification is mentioned as a possible subject of instruction[2]. The next two years were devoted to logic; the text-books being the *Summulae* and the commentary of Duns Scotus. The fourth year was given up to rhetoric: this meant certain parts of Aristotelian philosophy, as derived from Arabic sources.

Instruction in these subjects was given by the *cursory* lectures of students in their fifth, sixth, or seventh years of residence (which had to be delivered before nine in the morning or after noon); and by the *ordinary* lectures which every (regent) master was obliged to give for at least one year after taking his degree. All other lectures were termed *extraordinary.* Every lecture had to be given in the schools[3], and the university derived a considerable part of its scanty income from the rents taken from the lecturers. Gratuitous lectures were forbidden[4]. A statute of Urban V. in 1366 addressed to the university of Paris expressly forbad to students the use of benches or seats in lecture-rooms; this was probably held binding at Cambridge, and all students attending lectures were expected to sit or lie on straw scattered on the floor, as we know was the case in Paris. Only extraordinary lectures were permissible in the Long Vacation.

1 See the regulations of King's Hall, quoted in Mullinger, I. 253.

2 See Mullinger, I. 350.

3 A list of pictures of lectures in illuminated manuscripts is given in Lyte, 228.

4 *Cambridge documents*, I. 391; similar regulations existed at Oxford, *Munimenta academica*, 110, 129, 256, 279.

The lectures were either dictatory, or analytical, or dialectical[1]. The first or *nominatio ad pennam* consisted in dictating text-books, for few students possessed copies of any works except the *Summulae* and the *Sententiae:* the former being the standard work on logic, and the latter on theology. The second or analytical lecture was purely formal, and traditionally was never allowed to vary in any detail—an illustration of it is extant in the commentary by Aquinas on Aristotle's *Ethics*. The lecturer commenced with a general question; mentioned the principal divisions; took one of them and subdivided it; repeated this process over and over again till he got to the first sentence in that part of the work on which he was lecturing; he then expressed the result in several ways. Having finished this he started again from the beginning to get to his second sentence. No explanatory notes or allusions to other parts of the same work or to other authorities were permitted. These lectures were the resource of those masters who wished to get through their regency with as little trouble as possible, but for the credit of the mediæval students I am glad to say that they were not popular. Thirdly, there was the dialectical lecture, where each sentence, or some interpretation of it, was propounded as a question and defended against all objections, the arguments being thrown into the syllogistic form and of course expressed in Latin. Any student might be called on to take part in the discussion, and it thus prepared him for the ordeal through which he had subsequently to pass to obtain a degree. An illustration of this is extant in the *Quaestiones* of Buridanus.

To supplement the instruction given by the regents, three teachers (known as the Barnaby lecturers) were annually appointed by the university, at stipends of £3. 6*s*. 8*d*. a year, to lecture on Terence, logic, and philosophy[2]; and subsequently a fourth lectureship on the subjects of the quadrivium was

[1] See Mullinger, I. 359 et seq.; and Peacock, appendix A.

[2] See Peacock, appendix A, v.

created with a stipend of £4 a year[1]. These officers were regularly appointed till 1858, though for nearly three centuries they had given no lectures.

By the Lent term of his third year of residence a student was supposed to have read the subjects of the trivium, and he was then known as a *general sophister*. As such he had to dispute publicly in the schools four times; twice as a respondent to defend some thesis which he asserted, and twice as an opponent to attack those asserted by others. A bachelor presided over these discussions. The subject-matter of these acts in mediæval times was some scholastic question or a proposition taken from the *Sentences*. About the end of the fifteenth century religious questions, such as the interpretation of biblical texts, began to be introduced[2]. Some fifty or sixty years later the favorite subjects were drawn either from dogmatic theology (or possibly from philosophy). In the seventeenth century the questions were usually philosophical, but in the eighteenth century most of them were mathematical. Some of these are printed later. A complete list of the acts of any year would give a very fair idea of the prevalent studies.

After keeping his acts the sophister was examined by the university as to his character and academical standing, and if nothing was reported against him, presented himself as a *questionist* to be examined by the proctors and regents in the arts school. In general he had then to defend some question against the most practised logicians in the university—a somewhat severe ordeal. Stupid men propounded some irrefutable truism, but the ambitious student courted attack by affirming some paradox.

The influence of these acts, especially those for the higher degrees, was very considerable. Thus the brilliant declamation of Peter Ramus for his master's degree at Paris on the subject *Quaecumque ab Aristotele dicta essent commenticia esse* drew a crowded and critical audience, and the subsequent

[1] See *Statuta antiqua*, 136.

[2] Mullinger, I. 568.

discussion really affected the whole subsequent development of philosophy in Europe.

A candidate was never rejected, but reputation or contempt followed the popular verdict as to how he acquitted himself. The desirability of having on these occasions a numerous and friendly audience was so great that a man's friends not only came themselves, but used forcible means to bring in all passers-by. So considerable a nuisance did the practice become that a statute at Oxford is extant in which it is condemned under the penalty of excommunication and imprisonment[1].

This test having been passed the student obtained a *supplicat* to the senate from his hostel or college. He was then admitted as an incepting bachelor. This was not a degree, but it marked the transition to the studies and life of an undergraduate. The official account of the ceremony is sufficiently quaint to be worth quoting. On a day shortly before Ash-Wednesday about nine o'clock in the morning the bedells, each carrying his silver staff of office or *bacillarius* (from which, it has been suggested, the title of *bachelor* may possibly be derived[2]), "shall go to the College, House, Hall, or Hostel where the said Questionists be, and at their entry into the said House shall call and give warning in the midst of the Court with these words, *Alons, Alons, goe, Masters, goe, goe*; and then toll, or cause to be tolled the bell of the House to gather the Masters, Bachelors, Scholars, and Questionists together. And all the company in their habits and hoods being assembled, the Bedells shall go before the junior Questionist, and so all the rest in their order shall follow bareheaded, and then the Father, and after all, the Graduates and

[1] *Munimenta academica*, I. 247.

[2] See p. 208 of *University society in the eighteenth century*, by C. Wordsworth, Cambridge, 1874. The derivation usually given is from the Celtic *bach*, little, from which comes the old French *baceller*, to make love: but Prof. Skeat in his dictionary says that this is a bad guess, and in the supplement he repeats that the derivation is uncertain.

company of the said House, unto the common schools in due order. And when they do enter into the schools, one of the Bedells shall say, *noter mater* [*academia*], *bona nova, bona nova;* and then the Father being placed in the responsall's seat, and his children standing over against him in order, and the eldest standing in the hier hand and the rest in their order accordingly, the Bedell shall proclaim, if he have any thing to be proclaimed, and further say, *Reverende Pater, licebit tibi incipere, sedere, et cooperiri si placet.* That done, the Father shall enter his commendations[1] of his children, and propounding of his questions unto them, which the eldest shall first answer, and the rest in order. And when the Father has added his conclusion unto the questions, the Bedell shall bring them home in the same order as they went...and at the uttermost school door the Questionists shall turn them to the Father and the company and give them thanks for their coming with them[2]." But the regulations add that if the Father shall ask too hard questions or entrap his children into an argument "the Bedell shall knock him out," by which was meant knocking the door so loudly that nothing else could be heard.

At a later time the incepting bachelors were divided into classes, the higher classes being admitted to the title of bachelor a few weeks before the lower ones. The former correspond to the honour students of the present time, the latter to the poll men.

During the remainder of the Lent term the newly incepted bachelor was expected to spend every afternoon in the schools. In addition to the necessity of "disputing" with any regent who cared to come and test his abilities, he was required to preside at least nine times over the disputations which those who were studying the trivium were keeping, criticize the arguments used, and sum up or *determine* the whole discussion.

[1] At this point of the ceremony the candidates knelt, and the bedells are directed to pluck the hoods of the candidates over their faces, so that the blushes raised by their modesty may not be seen.

[2] Peacock, Appendix A, IV—VI.

Hence he was usually known as a *determiner*, and was said to stand in *quadragesima*.

There was a master of the schools whose business it was to keep order. But his task must have been very difficult, and apparently was generally beyond his powers; for we read that drinking, wrestling, cockfighting, and such like amusements were common. These "determinations" were regarded as a great opportunity for distinction, but the school was a rough one, and many students preferred to determine by proxy which was permissible[1].

It will be noticed that the quadragesimal disputations took place after Ash-Wednesday, and therefore after the admission of some or all the students to the title of bachelor. In early times it is believed that the inception took place even before the examination by the proctors.

The bachelor was supposed to devote the next three years to the study of the quadrivium; namely, arithmetic, geometry (including geography), music, and astronomy; and before he could proceed to the degree of master he had to make a declaration that he had studied these subjects. There was however no public test of his knowledge, and practically, unless he had a marked interest in them, he continued to devote his time to logic, metaphysics, or theology, which then afforded the only avenues to distinction.

I have already pointed out that a bachelor was expected to give cursory lectures, by which it may be added he earned some pocket-money. He was also required to be present at all public disputations of masters of arts unless expressly excused by the proctors, to keep three acts against a regent master, two acts against bachelors, and give one declamation.

It is usually said that most bachelors resided and in due course commenced master. That is true of scholars at the colleges who were obliged by statute to do so, but I suspect that most students at the hostels went down after their admission to the title of bachelor.

[1] See *Statuta antiqua*, 141.

At the end of the seventh year from his entry the student who had performed all these exercises could become a master. The degree itself or the formal ceremony of creation was given on the second Tuesday in July, called the day of commencement. On the previous evening certain exercises of inception, known as the vespers, were performed in the schools[1]. On the Tuesday morning the whole university met in Great St Mary's (which was fitted up for the occasion something like a theatre) at 7 A.M. to hear high mass. The supplicat for the degree was then presented. If this were passed the youngest regent present (or his proxy), known as the *praevaricator*, opened the proceedings with a speech in which any questions then affecting the university were discussed with considerable license. Next a doctor of divinity, acting as the "father," placed the *pileum* or cap (symbolical of a master's degree) on the head of the incepting master. The latter then defended a proposition taken from Aristotle, first against the prævaricator, and then against the youngest non-regent; finally the youngest doctor of divinity summed up the conclusion. Each successive inceptor went through a similar exercise.

Anthony Wood discovered a manuscript containing a few questions proposed at the similar congregation at Oxford. They apparently owe their preservation to the fact that the inceptor put the proposition into metrical form, which struck the audience as an ingenious conceit. I give one as a specimen of the kind of questions propounded. "Questio quinta ad quam respondebit quintus noster inceptor dominus Robertus Gloucestriæ, quæ de licentia duorum procuratorum et cum supportatione hujus venerabilis auditorii est diutius pertractanda, est in hac forma. *Utrum potentiarum imperatrix | celsa morum gubernatrix, | vis libera rationalis, | sit laureata dignitate | electionis consiliatae | ut Domina principalis.*"

[1] The students by immemorial custom were permitted to seize the new inceptor as he came out, and whether he liked it or no (and the extant references shew that he usually didn't) shave him in preparation for the morrow.

The subsequent ceremonies of inception are described at length in Peacock[1] and were chiefly formal. The incepting master was expected to make a present of either a gown or gloves to every officer of the university, and to give a dinner to all the regents, to which however he was allowed to ask his own friends. The cost of this must have been considerable. In the fourteenth century the universities of Paris, Oxford, and Cambridge passed identical statutes that no one should spend on his inception more than £41. 13*s.* 4*d.*, a sum which is equivalent to about £500 now, and must have been far above the means of most students[2]. Noblemen at Oxford and Cambridge were exempted from this restrictive rule[3].

A student could apparently plead poverty as an excuse for not fulfilling these duties, or could incept by proxy—the proxy receiving a degree too. The conditions under which this was allowed are not fully known.

These presents and the cost of the dinner were ultimately changed into a fee to the university chest. The difficulty of

[1] See Appendix A to Peacock's *Observations*.

[2] *Statuta antiqua*, 127. Mullinger, I. 357.

[3] I can quote the menu of one feast given by a wealthy inceptor, the cost of which must have far exceeded the statutable limit; but it owes its preservation to the fact that it was an exceptional case. The wealth of the host was fabulously large, and no conclusion can be drawn as to the usual practice. The "dinner" to which I refer was that given by George Nevill, the brother of the Earl of Warwick, on taking his master's degree in 1452. It lasted two days; on the first of which sixty, and on the second, two hundred dishes were served. The following is the bill of fare for the chief table, which in my ignorance of matters culinary I transcribe verbatim: a suttletee, the bore head and the bull; frumenty and venyson; fesant in brase; swan with chowdre; capon of grece; hernshew; poplar; custard royall; grant flanport desserted; leshe damask; frutor lumbent; a suttletee. The dishes served at the second table were viant in brase; crane in sawce; yong pocock; cony; pygeons; bytter; curlew; carcall; partrych; venyson baked; fryed meat in port; lesh lumbent; a frutor; a suttletee. At the third table were gely royall desserted; hanch of venson rosted; wodecoke; plover; knottys; styntis; quayles; larkys; quynces baked; viaunt in port; a frutor; lesh; a suttletee.

raising the money for these expenses was to some extent met by the university allowing the proctors to take jewels, manuscripts, or even clothes, as pledges. It would seem that the university sometimes made a bad bargain, for by a statute[1] of unknown date the proctors are forbidden to advance money on any books or manuscripts which are written on paper, but they are expressly allowed to continue to take vellum manuscripts as a security for fees. The new master was not permitted to exercise his functions until the term after that in which he incepted—a custom which still exists at Cambridge—but subject to that restriction he was obliged to reside and teach for at least one year, and was both entitled and obliged to charge a fee to those who attended his lectures. His duties were then at end, and if he went down he was tolerably sure of getting his livelihood, while his degree served as a license to lecture on the trivium and quadrivium in any university in Europe.

The genuine student, or the man who aimed at worldly success, generally proceeded to the doctor's degree in civil law, canon law, or theology; and in most colleges it was obligatory on a fellow to do so. A similar degree was also obtainable in medicine or music. No one could obtain the doctorate in any subject who did not really know it as it was then understood. These courses took from eight to ten years, and are too elaborate for me to describe here.

It was not uncommon for the new master to migrate to another university and take his doctorate there. Paris was especially thus favoured, and a mediæval scholar was rarely content if he had not spent a few years in the famous rue du fouarre. This migration facilitated the propagation of ideas, and served somewhat the same purpose as the multiplication of a book by printing at a later time.

If we were to judge solely by the statutes and ordinances of the university, this curriculum would seem to have been well designed as a general and elastic system of education. The scientific subjects of the quadrivium were however frequently

[1] See *Statuta antiqua*, 182.

neglected. This was partly due to the fact that they had practical applications, for the universities of Paris, Oxford, and Cambridge systematically discouraged all technical instruction, holding that a university education should be general and not technical. The chief reason for the neglect was however that no distinction could be obtained except in philosophy and transcendental theology. These subjects are interesting in themselves, and valuable as a branch of higher education, but experience seems to shew that only those who have already mastered some exact science are likely to derive benefit from their study. Be this as it may, it was not the belief of the schoolmen. They captured the mediæval universities, and there is a general consensus of opinion that the absence of fruitful work was mainly due to the fact that they controlled its studies and induced men to read philosophy before their opinions were sufficiently mature.

I should add that the popular idea that the schoolmen did nothing but dispute about questions such as how many angels could simultaneously dance on the point of a needle is grossly unjust. Besides discussing various questions which are still debated, they created the science of formal logic, and it is to them that the precision and flexibility of the Romance tongues is mainly due. No doubt some of their more foolish members said some foolish things, but to judge them by the propositions which Erasmus selected when he was attacking them and ridiculing their pretensions is manifestly unfair. It is said that in philosophy they settled nothing, but that was hardly their fault, for it is characteristic of the subject that no question is ever definitely settled. It must also be remembered that the schoolmen held that the value of a general education was to be tested by the methods used rather than the results attained.

The only subject that rivalled philosophy as a popular study was theology. It did not enter directly into the curriculum for the master's degree, but it involved the most burning questions of the day, and could not fail to excite general interest. The standard text-book for this was the work known as the *Sentences*[1]. This was a collection made by Peter Lombard, in 1150, of the opinions (*sententiae*) of the Fathers and other theologians on the most difficult points in the Christian belief. The logicians adopted it as a magazine of indisputable major premises, and created a large literature of deductions therefrom.

The period of transition.

The mediæval system of education was terminated by the royal injunctions of 1535, which forbad the teaching of the logic and metaphysics of the schoolmen, and in place thereof commanded the study of classical and biblical literature and of science. The subsequent rearrangements of the studies of the university were briefly as follows.

The first serious attempt to reorganize the studies of the university was embodied in the Edwardian code of 1549[2]. To check the presence of those who were merely schoolboys, it directed that for the future students (except those at Jesus College) should be required to have learnt the elements of Latin before coming into residence. The curriculum laid down was as follows. The freshman was to be first taught mathematics, as giving the best general training: this was to be followed by dialectics, and if desirable by philosophy: the whole forming the course for the bachelor's degree. The bachelor in his turn was expected to read perspective, astronomy, Greek, and the elements of philosophy before taking the master's degree. Finally, a resident master, after acting as regent for three years was expected to study law, medicine, or theology. These reforms represented the views of the moderate conservative party in the university, and the only objection expressed[3] was the very reasonable one that masters

[1] Mullinger, I. 59—63.

[2] Mullinger, II. 109—115.

[3] By Ascham: see p. 16 of *Original letters of eminent literary men* edited by Sir Henry Ellis, Camden Society, London, 1843.

should be at liberty to take the doctorate in any branch of literature or science that they pleased.

These statutes were replaced in 1557 by others, known as Cardinal Pole's; but the latter were repealed and the Edwardian (with a few minor alterations) re-enacted in 1559.

The period of transition was marked by the commencement of the professorial system of instruction. The mediæval plan of making every master lecture for at least one year was essentially bad; and in practice it had to be supplemented by the hostels and colleges. By the beginning of the sixteenth century it was generally admitted that this method was not adapted to the requirements of the university; and it was then proposed to endow professorships whereby it was hoped that the university would obtain for its students the best available teaching. The new system originated with the foundation in 1502[1] by the Lady Margaret of a chair of divinity; and in 1540 her grandson, Henry VIII., endowed the five regius professorships of divinity, law, physic, Hebrew, and Greek.

The age of transition was also contemporaneous with the establishment of the college system, as we know it. The early colleges were at first founded for a few fellows and scholars only. When however the insignificant little hall of God's House (which had been founded in 1439 and whose members never read beyond the trivium) was in 1505 enlarged and re-incorporated by Lady Margaret as Christ's College, a power was taken to admit pensioners, then called *convivae*, and at the same time the government was vested in the fellows as well as the master. These changes were introduced on the advice of Bishop Fisher, the confessor of Lady Margaret, to whom Cambridge is perhaps more indebted than to any other of its numerous and illustrious benefactors. A similar provision was inserted in the statutes of the other colleges which

[1] The earliest professorships founded at Oxford were those endowed by Henry VIII. in 1546. I believe professorships were established at Paris in the fifteenth century.

were shortly afterwards founded, viz. St John's, Buckingham (now known as Magdalene), Trinity, Emmanuel, and Sidney.

The colleges concerned themselves with the health, morals, and discipline of their students, as well as with their education. As soon as the college and university systems of tuition and discipline came into competition the latter broke down utterly[1]; and twenty years sufficed to change the university from one where nearly all the students were directly under the authority of the university to one where they were grouped in colleges, each college supervising the education and discipline of its students, subject of course to the general rules of the whole body of graduates by whom the final test of a proper education was applied before a degree was granted. The university imposed no exercises until a student's third year of residence and abandoned the duty of providing instruction for undergraduates to the colleges. It is easy to criticize the theory of the college system, but there can be no doubt that it at once met and still meets the general requirements of the nation at large.

The system of education under the Elizabethan statutes.

The period of transition in the studies of the university was brought to a close by the promulgation of the Elizabethan code of 1570, which remained almost intact till 1858. These statutes are memorable for the complete revolution which they effected in the constitution of the university, making it directly amenable to the influence of the crown and distinctly ecclesiastical in character. The manner in which these changes were

[1] Dr Caius had been educated under the old system, but when he returned in 1558 (to refound Gonville Hall) he found the collegiate system was firmly established. The history of the university which he wrote is thus particularly valuable, for he describes in detail exactly how the older system differed from that under which he then found himself living.

introduced is described later (see pp. 245–247). The curriculum was also recast[1]. Mathematics was again excluded from the trivium, and in lieu thereof undergraduates were directed to read rhetoric and logic; but the commissioners made no material alterations in the course for the master's degree. The power to interpret these statutes, and to arrange the times and details of all lectures and necessary exercises, was vested in the heads of colleges alone.

Although the subjects of education were changed the exercises for degrees, the manner of taking them, and the intervals between them were left substantially unaltered, save only that the conditions under which the exercises had to be performed were rigorously defined by statute, and no longer left to the discretion of the governing body of the university.

The statutable course for the degree of bachelor of arts was as follows[2]. An undergraduate was obliged to be a member of a college. After he had resided for three years[3], and had studied Greek, arithmetic, rhetoric, and logic, he was created a *general sophister* by his college. He then attended the incepting bachelors, comprising students one year senior to himself who were standing in quadragesima; and besides this read two theses, and kept at least two responsions and two opponencies under the regency of a master. At the end of his fourth[3] year he was examined by his college, and if approved presented as a *questionist*. In the week preceding Ash-Wednesday (or earlier in the same term) he was examined by the proctors (or by their deputies, the posers, subsequently termed moderators) and any other regents who wished to do so. A supplicat from the student's college was then presented, and if granted the undergraduate was admitted *ad respondendum quaestioni*. "I admit you," said the vice-chancellor, "to be bachelor of arts upon condition that you answer to your

[1] Mullinger, II. 232 *et seq.*

[2] Peacock, 8—10 *et seq.*

[3] The requisite residence was in practice shortened by reckoning the time from the term in which the name was put on the college boards.

questions: rise and give God thanks." The student then rose, crossed the senate-house, and knelt down to say "his private prayers." The ceremony of "entering the questions" took place immediately afterwards in the schools, the father or proctor asking a question from Aristotle's analytics. It was purely formal, and the bedells attended to "knock out" any one who began to argue. The questionist was admitted as a bachelor designate on Ash-Wednesday (or if not worthy of this was admitted a few weeks later). He then became a *determiner*, and after standing in quadragesima until the Thursday before Palm Sunday, the complete degree of *bachelor* was conferred by the proctors.

A candidate for the degree of master of arts was required to reside, to attend lectures, and to be present at all public acts kept by masters. Besides these he had to deliver one declamation, and to keep three respondencies against M.A. opponents, two respondencies against B.A. opponents, and six opponencies against B.A. respondents. The caput however in 1608 decided that residence should no longer be necessary for taking the master's degree. The decision was contrary to the statutes, but it only sanctioned a practice which had already become prevalent. The exercises and acts for that degree were thenceforth[1] reduced to a mere formality, so that the only real tests subsequently imposed by the university on its students were those immediately preceding and attending the admission to the bachelor's degree.

Like all immutable codes, which deal minutely with every detail of administration, the new statutes proved unworkable in some parts. It is doubtful if the performance of all the exercises and acts was ever enforced, and it was not long before some of the most important provisions of the new code were habitually and systematically neglected.

[1] I should add that in 1748 William Ridlington of Trinity Hall (B.A. 1739) who was then proctor, required the strict performance of the statutable exercises, and Christopher Anstey of King's was expelled for resisting the claim.

I come next to the method of giving instruction, which was usual during most of this period.

The professorial system was already well established. The regius chairs and others founded at a later time, brought eminent men to the university, and it would be difficult to overrate the influence thus exerted; but as a means of getting the best teaching suitable for the bulk of the students the scheme failed. In fact, the power of advancing the bounds of knowledge in any particular study and the art of expounding and teaching results that are already known are rarely united in the same person. The professors were generally selected for the first qualification. On the whole I think they were, in nearly all cases, the most eminent members of the university in their own departments; and if in the eighteenth century some of them not only did not teach but did very little to encourage advanced work, the fault is rather to be attributed to the age than to the system.

We must however recognize as a historical fact that till the end of the eighteenth century the professors did not—with a few exceptions, and notably of Newton—influence the intellectual life of the university as much as might have been reasonably expected, and they were generally glad to abandon nearly all teaching to the colleges.

Throughout the period in which the Elizabethan statutes were in force the college and tutorial systems of education were much as we now know them. I add in the following paragraphs a brief account of what the colleges expected from their students.

In the sixteenth century[1] an undergraduate was expected to rise at 4.30, after his private prayers (in a stated form) he went to chapel at 5.0. After service (and possibly breakfast) he adjourned to the hall, where he did exercises and attended lectures from six to nine. At nine the college lectures gene-

[1] This account is taken from the statutes of Trinity College: see Peacock, pp. 4—8. The statutes of 1552 and 1560 are printed as an appendix to the second volume of Mullinger's work.

rally ceased, and the great body of the students proceeded to the public schools, either to hear lectures, or to listen to, or take part in the public disputations which were requisite for the degree of bachelor or master. Dinner was served at eleven, and at one o'clock the students returned to their attendance on the declamations and exercises in the schools. From three until six in the afternoon they were at liberty to pursue their amusements or their private studies: at six o'clock they supped in the college-hall and immediately afterwards retired to their chambers. There was no evening service in the college chapels on ordinary days until the reign of James I. Whether most students lived up to this ideal is doubtful: some certainly did not.

As time went on the average age at entrance rose from about sixteen in the sixteenth century to seventeen or eighteen in the seventeenth, and to eighteen or nineteen in the eighteenth century. The hours also gradually got later, and the strictness of the regulations was somewhat relaxed. At the beginning of the eighteenth century the "college day began with morning chapel, usually at six. Breakfast was not a regular meal, but it was often taken at a coffee-house where the London newspapers could be read. Morning lectures began at seven or eight in the college-hall. Tables were set apart for different subjects. At 'the logick table' one lecturer is expounding Duncan's treatise, while another, at 'the ethick table' is interpreting Puffendorf on the duty of a man and a citizen; classics and mathematics engage other groups. The usual college dinner-hour which had long been 11 a.m., had advanced before 1720 to noon. The afternoon disputations in the schools often drew large audiences to hear respondent and opponent discuss such themes as 'natural philosophy does not tend to atheism,' or 'matter cannot think.' Evening chapel was usually at five; a slight supper was provided in hall at seven or eight[1]", or in summer even later. Sometimes after supper acts (preparatory to those in the schools) were kept:

[1] See Jebb's *Life of Bentley*, p. 88.

the origin of the college fees for those degrees is the remuneration paid to the M.A.'s who presided at these intramural exercises. At other times plays were then performed in hall, and once a week a *viva voce* examination (of course in Latin) was held. Some of the tutors also gave evening lectures in their rooms.

In the sixteenth and seventeenth centuries the educational work of the university was mainly performed by the college tutors. It was at first usual to allow men to choose each his own tutor according to the subject he wished to read, and to allow any fellow or the master to take pupils[1]; but the administrative and disciplinary difficulties connected with such a scheme proved insuperable, while it was found to be almost impossible for a corporation to prevent an inefficient fellow from taking pupils. The number of tutors was therefore limited, but it was still assumed that a tutor was able to give to every man all the instruction he required. Of course this universal knowledge was not generally possessed, and towards the beginning of the eighteenth century we hear of other teachers who were ready to give instruction in all the mathematical subjects required by the university.

There can be no question that some members of the university had given such private instruction in earlier times. I should however say that the difference between the mediæval system of coaching and that which sprang up in the eighteenth century was that the former was resorted to either by students who were backward and wanted special assistance, or by those who wished to specialize and went to specialists, while the latter was used by those who desired to master the maximum number of subjects in the minimum time with a view to taking as high a place in the tripos as possible. As soon as that examination, with its strictly defined order of merit, became the sole avenue to a degree coaching became usual and perhaps

[1] On the former tutorial system see e.g. the *Scholae academicae*, 259 *et seq.*; and also vol. ii., pp. 438—9 of Todhunter's *Life of Whewell*, London, 1876.

inevitable, for a high place in the tripos was not only the chief university distinction, but had a considerable pecuniary value.

There is no doubt that mathematics is most efficiently taught either by private instruction, or by lectures supplemented by private instruction. Every part of it has to be read in a tolerably well-defined sequence, and with the varying abilities and knowledge of men this requires a certain amount of individual assistance which cannot be given in a large lecture. Most of the tutors and professors of the eighteenth century neglected this fact. Indeed the professors, taken as a whole, made no effort to influence the teaching of the university, while the majority of the college tutors of that time were not sorry to be relieved of the most laborious part of their work. On the other hand, the instruction given by the coaches was both thorough and individual; while as men were free to choose their own private tutor, inefficient teachers were rare. Of course where the examination included a very large subject, such as a book of the *Principia*, that subject had to be taught by means of an analysis, and such analyses and manuscripts containing matter not incorporated into text-books were and are in constant circulation in the university.

The result of the movement was that the whole instruction of the bulk of the more advanced students (in mathematics) passed into the hands of a few men who were independent both of the university and of the colleges—a fact which seems to be as puzzling as it is inexplicable to foreign observers.

I am satisfied that the system originated in the eighteenth century, but I have found it very difficult to arrive at any definite facts or dates. In particular I am not clear how far the "pupil-mongers" of the beginning of that century, such as Laughton, are to be regarded as private tutors or not. I suspect that they were college lecturers who threw their lectures open to the university, but supplemented them by additional assistance for which they were paid a private fee.

The earliest indisputable reference to a coach across which I have come is in the life[1] of William Paley of Christ's. His "private tutor" was Wilson of Peterhouse (see p. 102), by whom "he was recommended to Mr Thorp [Robert Thorp, of Peterhouse, B.A. 1758, and afterwards archdeacon of Northumberland] who was at that time of eminent use to young men in preparing them for the senate-house examination and peculiarly successful. One young man of no shining reputation with the assistance of Mr Thorp's tuition had stood at the head of wranglers." Thorp—to cut a long story short—consented to coach Paley, and brought him out as senior in 1763. A grace passed by the senate in 1781 commences with a preamble in which it is stated that almost all sophs then resorted to private tuition.

At that time the moderators in the tripos often prepared pupils for the examination they were about to conduct. Various graces[2] of the senate were passed from 1777 onwards to stop this custom. At a later period different attempts were made to prevent private tutors from acting as examiners, but all such legislation broke down in practice.

Even non-residents acquired a reputation as successful coaches. Thus John Dawson, a medical practitioner at Sedbergh (born in January, 1734, and died in September, 1820), regularly prepared pupils for Cambridge, and read with them in the long vacation. At least eleven of the senior wranglers between 1781 and 1800 are known to have studied under him, but the names of his pupils cannot in general be now determined.

During the first three-quarters of the present century (i.e. beyond the point to which my history extends) nearly

[1] See p. 29 of his life by E. Paley, London, 1838. William Paley was the author of the well known *View of the evidences of Christianity*, first published in 1794: he was born in 1743, and died in 1803.

[2] A list of them is given in chap. III. section 3 of Whewell's *Of a liberal education*, second edition, London, 1850. See also the *Scholae academicae* pp. 260—261.

every[1] mathematical student read with a private tutor. So universal was the practice that William Hopkins (who was born in 1805, graduated as seventh wrangler in 1827, and died in 1866) was able, in 1849, to say that since his degree he had had among his pupils nearly two hundred wranglers, of whom 17 had been senior and 44 in one of the first three places. So again at the recent presentation of his portrait to Dr Routh by his old pupils it was remarked that he had directed the undergraduate mathematical education of nearly all the younger Cambridge mathematicians of the present time. Thus in the thirty-one years from 1858 to 1888 he had had no less than 631 pupils, most of whom had been wranglers, and 27 of whom had been senior wranglers.

Private tuition in other subjects became for a short time usual, but with the recent developments and improvements in college teaching by the aid of a large staff of teachers in addition to the tutors, the necessity for coaching has gradually disappeared—at any rate in subjects other than mathematics. Whether in that subject it is possible to give all the requisite teaching by college lectures without sacrificing the advantages of order of merit in the tripos is one of the problems of the present time.

[1] There were exceptions; thus G. Pryme, who was sixth wrangler in 1803, writes in his *Reminiscences* (p. 48) that coaching was not really necessary, and that he found college lectures sufficient.

CHAPTER IX.

THE EXERCISES IN THE SCHOOLS[1].

I PURPOSE now to give an account of the scholastic acts to which so many references were made in the last chapter, and to illustrate their form by reproducing one on a mathematical subject.

I have already enumerated the subjects of instruction enjoined by the Elizabethan statutes, and it is certain that it was intended that the scholastic disputations should be kept on philosophical questions drawn from that curriculum.

The statutes however had hardly received the royal assent before the philosophy of Ramus (see p. 14) became dominant in the university; and the discussions were tinged by his views. About 1650 the tenets of the Baconian and Cartesian[2] systems of philosophy became the favourite subjects in the schools of the university. Some fifty years later they were displaced by subjects drawn from the Newtonian philosophy, and thenceforth it was usual to keep some of the disputations on mathematical subjects; though it always remained the general custom to

[1] The substance of this chapter is reprinted from my *Origin and history of the mathematical tripos*, Cambridge, 1880. The materials for that were mainly taken from *Of a liberal education*, by W. Whewell, Cambridge, 1848, and the *Scholae academicae*, by C. Wordsworth, Cambridge, 1877.

[2] I think there can be no doubt that the Cartesian philosophy was read: Whewell, however, always maintained the contrary, but in this opinion he was singular.

propound at least one philosophical question, which was frequently taken from Locke's *Essay*. In 1750 it was decided in Cumberland's case that it was not necessary for a candidate to offer any except mathematical subjects.

The earliest list with which I am acquainted of questions kept in the schools is contained in the *Disputationum academicarum formulae* by R. F., published in 1638. A list of questions on philosophy in common use during the early years of the eighteenth century was published in 1735 by Thomas Johnson, who was a fellow of Magdalene College and master at Eton.

The procedure seems to have remained substantially unaltered from the thirteenth to the nineteenth centuries, and it is probable that the following account taken from the records of the eighteenth century would only differ in details from the description of a similar exercise kept in the middle ages.

The disputation commenced by the candidate known as the *act* or *respondent* proposing three propositions [in the middle ages he only proposed one] on one of which he read a thesis. Against this other students known as *opponents* had then to argue. The discussions were presided over by the moderators [or before 1680 by the proctors, or their deputies the posers], who moderated the discussion and awarded praise or blame as the case might require. The discussions were always carried on in Latin and in syllogistic form.

In the eighteenth century, when the system had crystallized into a rigid form, it was the invariable custom to have in the sophs's schools three opponents to each respondent. Of these the first, who took the lead in the discussion, was expected to urge five objections against the first of the propositions laid down by the respondent, three against the second, and one against the third. The respondent replied to each in turn, and when an argument had been disposed of, the moderator called for the next by saying *Probes aliter*. When the disputation had continued long enough the opponent was dismissed with some such phrase as *Bene disputasti*. The second opponent followed, and urged three objections against the first proposition and one against each of the others. His place was then taken by the third opponent, of whom but one argument against each question was required. If a candidate failed utterly he was dismissed with the order *Descendas*, which was equivalent to a modern pluck. Such cases were extremely rare. Finally, the respondent was examined by the moderator, and according as he acquitted himself was released with some suitable remark.

The following is a more detailed account of the procedure in the eighteenth century. By that time all the exercises subsequent to the admission to the degree of bachelor had become reduced to a mere formality; but every student (unless he intended to proceed in civil law, or was a fellow-commoner) had in the course of his third year of residence to keep one or more disputations in the sophs's schools.

At the beginning of the Lent term the moderators (or, before 1680, the proctors) applied to the tutors of the different colleges for lists of the candidates for the next year. An undergraduate had no right to present himself, and several cases are mentioned in which permission to keep exercises in the schools was refused to students who were not likely to do credit to the college. To see if this were the case it was usual for the college authorities to examine their students before the latter were allowed to keep an act in public, and to prepare them for it by mock exercises in the college hall. The college fee for students taking a bachelor's or master's degree was, as I have already said, originally imposed to cover the cost of this preliminary examination and preparation.

The lists sent by the college tutors were supplemented by memoranda such as 'reading man,' 'non-reading man,' &c., and guided by these remarks and the general reputation of the students the moderators fixed on those who should keep the acts and opponencies. The expectant wranglers were generally chosen to be the respondents, they and the senior optimes were reserved for the first and second opponencies (on whom the brunt of the discussion fell), and the third opponencies were given to those who were expected to take a poll degree, the appearance of the latter in the schools being often little more than nominal.

By a happy accident the private list of Moore Meredyth, of Trinity (B.A. 1736), who was one of the proctors for 1752 has been preserved, and is now in the university registry. It contains altogether the names of seventy-seven students[1]. Of these twelve are placed first in a class by themselves headed by the letter *R*, which means that they were selected to be respondents. Fourteen are put next by themselves in another division marked *O*, and these men were most likely chosen to keep first opponencies. The names of those who were not expected to take honours form a third list. The names in each set begin with the Trinity men, and those from the other colleges follow.

From the list which the moderators had thus drawn up of the candidates, and some three weeks before any particular respondent had to keep an act, he received a notice from the proctors calling on him to propose for their approval three subjects for discussion. In practice he was allowed to choose any questions taken from the traditional subjects of examination, and to select the one in support of which he should read his thesis. So important was the work of preparation that even a college dean relented somewhat of his sternness, and the student was permitted to take out a *dormiat*, and thus excused from morning chapels was able to concentrate all his attention on the approaching contest. One of his first duties was to make the acquaintance of his opponents, inform them on which of the three subjects he intended to read his thesis, and arrange other details of the contest. In earlier times the opponents had no such assistance. The opponents in a similar way arranged amongst themselves the order and plan of their arguments.

The disputation began about three o'clock. As soon as the moderator had taken his seat he said *Ascendat dominus*

[1] *Scholae academicae*, pp. 363, 364.

respondens, and thereupon the respondent walked up into a sort of desk facing the moderator. The exercise commenced by his reading a Latin thesis, which lasted about ten minutes, in support of one of his propositions: this essay was afterwards given to the moderators. As soon as it was finished the moderator, turning to the first opponent, said *Ascendat opponentium primus.* The latter then entered a box below or by the side of the moderator and facing the respondent. He opposed the proposition laid down in the thesis in five arguments, the second question in three, and the third in one. Every argument was put into the form of a hypothetical syllogism and ran as follows. *Major premise*: If *A* is *B* (the *antecedentia*) *C* is *D* (the *consequens*, or more generally but inaccurately spoken of as the *consequentia*). *Minor premise*: But *A* is *B*. *Conclusion*: Therefore *C* is *D* (the *consequentia*). The respondent denied any step in this that was not clear, generally admitting that *A* was *B*, but alleging that it did not follow that *C* was *D*. The opponent then explained how he maintained his objection, and this process was continually repeated until he had fairly stated his case, when the respondent replied; and the discussion was then carried on until the moderator stopped it by saying to the opponent *Probes aliter.* After the eighth argument the first opponent was sent down with some compliment such as *Domine opponens, bene disputasti,* or *optime disputasti*, or even *optime quidem disputasti.* It is from this use of the word that the terms senior optime and junior optime are derived. As soon as the first opponent had finished, the second opponent followed and urged three objections against the first proposition and one against each of the others. His place was then taken by the third opponent, of whom but one argument against each question was required. Finally, the respondent was examined by the presiding moderator, and according as he did badly or well was released with the remark *Tu autem, domine respondens, bene* (or *satis*, or *satis et bene*) *disputasti*, or even *satis et optime quidem et in thesi et in disputationibus tuo officio functus es*, or sometimes with the highest compliment of all, *summo ingenii acumine disputasti.*

In general *optime quidem* was the highest praise expected, but towards the close of the eighteenth century Lax introduced the custom of giving elaborate compliments, much to the disgust of some of the older members of the university. An order to quit the desk was equivalent to rejection, but the power was very rarely used.

A copy of the thesis read on Feb. 25, 1782, by John Addison Carr of Jesus for his act is in the library of Trinity[1], it is apparently the original manuscript handed to the moderators at the close of the disputation. The manuscript begins

Q[uæstiones] S[unt]
Recte statuit Newtonus in tertia sua sectione.
Recte statuit Emersonus de motu projectilium.
Origo mali moralis solvi potest salvis Dei attributis.
De postrema.

Then follows an essay on the third question ; and on the last page of the manuscript there is a memorandum

Carr, coll. Jes. Resp. Feb. 25, 1782.
Bere, Sid. coll., Opp. 1^mus^.
Cragg, S.S. Trinitatis, Opp. 2us.
Newcome, coll. Regin., Opp. 3us.

Finally at the bottom is the signature of the presiding moderator *Littlehales Mod^r^. Coll. Johann.* which he affixed at the conclusion of the act. The essay covers some eight and a half closely written pages of a foolscap quarto note-book, and is not worth quoting. In the tripos list of 1783, Carr came out as eleventh senior optime, Bere as ninth senior optime, Cragg as sixth junior optime (i.e. last but two), and Newcome as twelfth wrangler.

On the results of these discussions the final list of those qualified to receive degrees was prepared. The order of this list in early times had been settled according to the discretion

[1] The Challis manuscripts, III. 59.

of the proctors and moderators, and every candidate before presenting himself took an oath that he would abide by their decision. The list was not arranged strictly in order of merit, because the proctors could insert names anywhere in it; but except for these honorary distinctions, the recipients of which were called proctors's or honorary optimes, it probably fairly represented the merits of the candidates. The names of those who received these honorary degrees subsequent to 1747 are struck out from the lists given in all the calendars issued subsequent to 1799. It is only in exceptional cases that we are acquainted with the true order for the earlier tripos lists, but in a few cases the relative positions of the candidates are known; for example, in 1680 Bentley came out third though he was put down as sixth in the list of wranglers. By the beginning of the eighteenth century this power had apparently become restricted to the right reserved to the vice-chancellor, the senior regent, and each proctor to place in the list one candidate anywhere he liked—a right which continued to exist till 1827, though it was not exercised after 1797.

Subject to the granting of these honorary degrees, this final list was arranged in order of merit into three classes, consisting of (i) the wranglers and senior optimes; (ii) the junior optimes who had passed respectably but had not distinguished themselves; and (iii) *οἱ πολλοί*, or the poll men. The first class included those bachelors *quibus sua reservatur senioritas comitiis prioribus*: they received their degrees on Ash-Wednesday, taking seniority according to their order on the list. The two other classes received their degrees a few weeks later.

The order as determined by the performance of these acts seems to have been accurately foreshadowed by the preliminary lists framed by the moderators. Thus the tripos list for 1753 shews that all the undergraduates selected to be respondents became wranglers. Of the first opponents, three (probably personal friends of the moderators) received honorary optime degrees as second, third, and fourth wranglers respectively; four obtained a place in the first class by their own merits; and the rest appear as senior optimes—one, who was ill, receiving it as an honorary degree. The book lay before the moderators during the discussions, and if any third opponent shewed unexpected skill in the acts his name was marked, and transferred from the seventh or eighth class comprising the poll men to the fifth or sixth which contained the expectant junior optimes. In the list of 1752 sixteen names are thus crossed out, and these form the third class of that tripos. The rest of the candidates, thirty-five in number, together with seven others who kept no acts (at any rate before the moderators) form the poll list for that year.

At a later time, as we shall see in the next chapter, the acts were only used as a means of arranging the men into four groups, namely, those expected to be wranglers, senior optimes, junior optimes, and poll men respectively; and the order in each group was determined by the senate-house examination, in which a different set of papers was given to each group. Finally, a means of passing from one group to another by means of the senate-house examination was devised. Thenceforth the acts ceased to be of the same importance, though they still afforded a test by which public opinion as to the abilities of men was largely influenced.

The moderators's book for 1778 has been preserved and is in the library of Trinity. It may be interesting if I describe briefly the way in which it is arranged. Each page is dated, and contains a list of the three subjects proposed for that day together with the names of the respondent and the three opponents. Of the three questions proposed by each respondent the first was invariably on a mathematical subject, and with one exception was always taken from Newton. In all but ten cases the second was also on some mathematical question. The last was on some point in moral philosophy.

According as the acts were well kept or not the moderators marked the names of the candidates. Very good performances were rewarded with the mark $A+$, A, or $A-$; good performances with $E+$, E, or $E-$; fair performances with $a+$, a, or

$a-$; and indifferent ones with $e+$ or e. It was on these marks that the subsequent "classes" were drawn up.

Between Feb. 3 and July 2 sixty-six exercises in all were kept, each of course involving four candidates: between Oct. 26 and Dec. 11 thirty were kept. Three acts were stopped when only half finished because the book of statutes (without the presence of which a moderator had no power) was sent for by the proctors to consult at a congregation[1]. Two or three others are included in the book but are cancelled; most of them I gather because of some irregularity, but one because the selected respondent had died.

Altogether 112 students of that year presented themselves for the bachelor's degree, but they did not all appear in the schools. Of the honour candidates, forty-seven in number, one kept two acts, another kept three, and three kept four; all the rest kept five, six, or seven acts. Five honorary optime degrees were also given. There were sixty poll men: of these thirty-seven presented themselves at the proper time and formed the first list, all save eight of these having kept one or more acts. Eight bye-term men received their degrees as *baccalaurei ad baptistam* in the following Michaelmas term, and eight more as *baccalaurei ad diem cinerum* on Ash-Wednesday or "dunces's day." It was not usual for the

[1] Thus W. Chafin of Emmanuel, describing his act kept in 1752, says that he had got off tolerably well against W. Disney of Trinity, who was his first opponent, but that W. Craven of St John's "brought an argument against me fraught with fluxions; of which I knew very little and was therefore at a nonplus, and should in one minute have been exposed, had not at that instant the esquire bedell entered the schools and demanded the book which the moderator carries with him, and is the badge of his office. A convocation was that afternoon held in the senate-house, and on some demur that happened, it was found requisite to inspect this book, which was immediately delivered, and the moderator's authority stopped for that day, and we were all dismissed; and it was the happiest and most grateful moment of my life, for I was saved from imminent disgrace, and it was the last exercise that I had to keep in the schools." (From the *Gentleman's magazine* for January, 1818; quoted on pp. 29, 30 of the *Scholae academicae.*)

moderators to preside over the acts of bye-term men, and the exercises of these sixteen men do not therefore appear in this book. Of the remaining candidates two were "plucked" outright, four took a poll degree in the following year, and one candidate died during his questionist's year.

The senior wrangler of the year was Thomas Jones of Trinity, whose reputation, if we may believe tradition, was so well established that his attendance at the senate-house examination was excused by the moderators. Of course this did not prevent his position as senior being challenged (in the manner described on p. 200) if any candidate thought himself badly used. Jones had "coached" the second wrangler in his own year. He was afterwards tutor of Trinity, and one of the most influential members of the university at the end of the last century.

No detailed records of these disputations prior to the eighteenth century now exist. The official accounts by the proctors and moderators were usually destroyed as soon as the men were admitted to their degrees, and it is only by accident that the two from which I have made quotations above have been preserved.

The only verbatim reports (with which I am acquainted) of any disputations actually kept are of some which took place between 1780 and 1790. These are contained in a small manuscript now in the library of Caius College. One of them, by the kindness of that society, I was able to insert in my *Origin and history of the mathematical tripos*, published at Cambridge in 1880, and I here reproduce it. The manuscript consists of rough notes of exercises performed in the schools, with the addition of suggested objections to the questions most usually chosen by the respondents. Many of the arguments are crossed out as being obviously untenable, while several of the pages are torn and defaced, presenting much the same appearance as a copy book of an ordinary schoolboy would if it were preserved in some library as the sole specimen of its kind. Altogether the manuscript contains the whole or portions of twenty-three distinct disputations.

The conversational parts (i.e. the real discussions) are omitted throughout—indeed it was useless to take notes of these, since the debate was not likely to take exactly the same turn on any subsequent occasion—and the collection should therefore be regarded as an analysis of the arguments brought forward rather than as giving the actual disputations.

The discussion to which I alluded and which I here quote as an illustration of the form of these scholastic exercises was kept on Feb. 20, 1784, by Joshua Watson of Sidney, as first opponent, against the questions proposed by William Sewell of Christ's. The report of it is one of the fullest of those preserved in the book, and it seems also a good example both of the nature of the objections raised, and the form in which they were urged. In reference to the former, it is only fair to remember that the opponent had in general to deny a proposition which he knew perfectly well was true, and which the respondent had usually chosen because it was very difficult to controvert. In reference to the latter, the minor premise has been omitted from the manuscript in all save one of the disputations, but I have ventured to replace it and to add such other technical phrases as were always used. I have only to add that those portions which are not in the original are printed in square brackets: and that wherever the mark † is placed, there are pencil notes explaining how the conclusion is deduced; but time has rendered these so illegible that it is impossible to decipher them with certainty. The Latin is that of the schools, and I reprint it as it stands in the original.

The propositions were (i) Solis parallaxis ope Veneris intra solem conspiciendæ a methodo Halleii recte determinari potest; (ii) Recte statuit Newtonus in tertia sua sectione libri primi; (iii) Diversis sensibus non ingrediuntur ideæ communes.

After Sewell had read an essay on the first of these questions, the discussion began as follows.

Moderator. [Ascendat dominus opponentium primus.]

Opponent. [Probo] contra primam [quæstionem]. Si asserat Halleius Venerem cum Soli sit proxima Londini visam a centro Solis quatuor minutis primis distare, cadit quæstio. [Sed asserit Halleius Venerem cum Soli sit proxima Londini visam a centro Solis quatuor minutis primis distare. Ergo cadit quæstio.]

Respondent. [Concedo antecedentiam et nego consequentiam.]

Opp. [Probo consequentiam.] Si in schemate posuit semitam Veneris ad os Gangeticum quatuor etiam minutis primis distare, valet consequentia [Sed in schemate posuit semitam Veneris ad os Gangeticum quatuor etiam minutis primis distare. Ergo valet consequentia.]

Resp. [Concedo antecedentiam et nego consequentiam.]

Opp. [Iterum probo consequentiam.] Si spectatoribus positis in diversis parallelis latitudinis non eadem appareat distantia atque igitur non licet eandem visibilem sumere distantiam in hisce duobus locis valent consequentia et argumentum. [Sed spectatoribus positis in diversis parallelis latitudinis non eadem apparet distantia atque non licet eandem visibilem sumere distantiam in hisce duobus locis. Ergo valent consequentia et argumentum.]

The conclusion *valet argumentum* meant that the opponent considered that he had fairly stated his case, and here therefore ought to follow first the respondent's exposition of the fallacy in the opponent's argument, and then the opponent's answer sustaining his objection to the original proposition given above. As soon as each had fairly stated and illustrated his case or the discussion began to degenerate into an interchange of personalities, the moderator turning to the opponent said *Probes aliter*, and a fresh argument was accordingly begun. All these steps are missing in the manuscript.

The remaining seven arguments of the opponent were as follows.

Opp. [Probo] aliter [contra primam]. Si in figura Halleiana centrum Solis correspondeat cum loco spectatoris in Tellure, cadit quæstio. [Sed in figura Halleiana centrum Solis correspondet cum loco spectatoris in Tellure. Ergo cadit quæstio.]†

Resp. [Concedo antecedentiam et nego consequentiam.]

Opp. [Probo consequentiam.] Si locus centri Solis a vero centro amoti ob motum spectatoris fit curva linea, valet consequentia. [Sed locus centri Solis a vero centro amoti ob motum spectatoris fit curva linea. Ergo valet consequentia.]

Resp. [Concedo antecedentiam et nego consequentiam.]

Opp. [Iterum probo consequentiam.] Si composito motu Veneris uniformi in recta linea cum motu Solari in curva linea fit semita Veneris in disco Solis curva linea, valet consequentia. [Sed composito motu Veneris uniformi in recta linea cum motu Solari in curva linea fit semita Veneris in disco Solis curva linea. Ergo valet consequentia.]

Resp. [Concedo antecedentiam et nego consequentiam.]

Opp. [Iterum probo consequentiam.] Si longitudo hujusce lineæ non recte determinari potest, valent consequentia et argumentum. [Sed longitudo hujusce lineæ non recte determinari potest. Ergo valent consequentia et argumentum.]

The next argument against the first proposition ran as follows.

Opp. [Probo] aliter [contra primam]. Si spectatori ad os Gangeticum posito ob terræ motum motui Veneris contrarium contrahatur transitus tempus integrum, cadit quæstio. [Sed spectatori ad os Gangeticum posito ob terræ motum motui Veneris contrarium contrahitur transitus tempus integrum. Ergo cadit quæstio.]

Resp. [Concedo antecedentiam et nego consequentiam.]

Opp. [Iterum probo consequentiam.] Si assumat Halleius contractionem hanc duodecim minutis primis temporis æqualem, et deinde huic hypothesi insistendo eidem tempori æqualem probat, valent consequentia et argumentum. [Sed assumat Halleius contractionem hanc duodecim minutis primis temporis æqualem, et deinde huic hypothesi insistendo eidem tempori æqualem probat. Ergo valent consequentia et argumentum.]

The fourth objection to the first proposition was as follows.

Opp. [Probo] aliter [contra primam]. Si posuit Halleius eandem visibilem semitam Veneris per discum Solarem ad os Gangeticum et portum Nelsoni, et hanc semitam dividat in æqualia horaria spatia, cadit quæstio. [Sed Halleius posuit eandem visibilem semitam Veneris per discum Solarem ad os Gangeticum et portum Nelsoni, et hanc semitam dividit in æqualia horaria spatia. Ergo cadit quæstio.]

Resp. [Concedo antecedentiam et nego consequentiam.]

Opp. [Probo consequentiam.] Si motus horarius Veneris acceleratur vel retardatur per motum totum spectatoris in medio transitu, quo magis autem distat, minus acceleratur vel retardatur, valet consequentia. [Sed motus horarius Veneris acceleratur vel retardatur per motum totum spectatoris in medio transitu, quo magis autem distat, minus acceleratur vel retardatur. Ergo valet consequentia.]

Resp. [Concedo antecedentiam, et nego consequentiam.]

Opp. [Iterum probo consequentiam.] Si igitur ob motum Veneris acceleratum ad os Gangeticum et retardatum ad portum Nelsoni hi motus non debent repræsentari per idem spatium, valent consequentia et argumentum. [Sed ob motum Veneris acceleratum ad os Gangeticum et retardatum ad portum Nelsoni hi motus non debent repræsentari per idem spatium. Ergo valent consequentia et argumentum.]

The last argument against the first question was as follows.

Opp. [Probo] aliter [contra primam]. Si secundum constructionem Halleianam spectatori ad portum Nelsoni, posito tempore extensionis majore, major etiam fit transitus duratio, cadit quæstio. [Sed secundum constructionem Halleianam spectatori ad portum Nelsoni, posito tempore extensionis majore, major fit transitus duratio. Ergo cadit quæstio.]†

Resp. [Concedo antecedentiam et nego consequentiam.]

Opp. [Probo consequentiam.] Si secundum eandem constructionem posito quod spectatori ad os Gangeticum tempus contractionis majus sit duodecim minutis primis, evadat tempus durationis majus etiam, valet consequentia. [Sed secundum eandem constructionem posito quod spectatori ad os Gangeticum tempus contractionis majus est duodecim minutis primis, et evadit tempus durationis majus etiam. Ergo valet consequentia.]†

Resp. [Concedo antecedentiam et nego consequentiam.]

Opp. [Iterum probo consequentiam.] Si hæ duæ conclusiones inter se pugnent, valent consequentia et argumentum. [Sed hæ duæ conclusiones inter se pugnant. Ergo valent consequentia et argumentum.]

The opponent then proceeded to attack the second proposition, and his first objection to it was as follows.

Opp. [Probo] contra secundam [quæstionem]. Si vis in parabola ad infinitam distantiam sit infinitesimalis secundi ordinis, cadit quæstio. [Sed ad infinitam distantiam vis in parabola est infinitesimalis secundi ordinis. Ergo cadit quæstio.]

Resp. [Concedo antecedentiam et nego consequentiam.]

Opp. [Probo consequentiam.] Si vis sit u^4 igiturque ad infinitam distantiam sit infinitesimalis quarti ordinis, valent consequentia et argumentum. (*The manuscript here is almost unintelligible.*) [Sed vis est u^4 igiturque ad infinitam distantiam est infinitesimalis quarti ordinis. Ergo valent consequentia et argumentum.]

The second objection to this question was as follows.

Mod. [Probes aliter.]

Opp. [Probo] aliter [contra secundam]. Si velocitates ad extremitates axium minorum diversarum ellipsium quarum latera recta æquantur sint inter se inverse ut axes minores, cadit quæstio. [Sed velocitates ad extremitates axium minorum diversarum ellipsium quarum latera recta æquantur sunt inter se inverse ut axes minores. Ergo cadit quæstio.]

Resp. [Concedo antecedentiam et nego consequentiam.]

Opp. [Probo consequentiam.] Si locus extremitatum omnium axium minorum sit parabola, valet consequentia. [Sed locus extremitatum omnium axium minorum est parabola. Ergo valet consequentiam.]

Resp. [Concedo antecedentiam et nego consequentiam.]

Opp. [Iterum probo consequentiam.] Si velocitas corporis revolventis in ista parabola sit ad velocitatem ad mediam distantiam correspondentis ellipseos ut $\sqrt{2} : 1$, valet consequentia. [Sed velocitas corporis revolventis in ista parabola est ad velocitatem ad mediam distantiam correspondentis ellipseos ut $\sqrt{2} : 1$. Ergo valet consequentia.]

Resp. [Concedo antecedentiam et nego consequentiam.]

Opp. [Iterum probo consequentiam.] Si velocitas in parabola sit inverse ut ordinata, valent consequentia et argumentum. [Sed velocitas in parabola est inverse ut ordinata. Ergo valent consequentia et argumentum.]

The argument against the third proposition was as follows.

Mod. [Probes aliter.]

Opp. [Probo] contra tertiam [quæstionem]. Aut cadit tua quæstio aut non possibile est hominem ab ineunte ætate cæcum et jam adultum visum recipientem visu dignoscere posse id quod tangendo prius solummodo dignoscebat. Sed possibile [est hominem ab ineunte ætate cæcum et jam adultum visum recipientem visu dignoscere posse id quod tangendo prius solummodo dignoscebat. Ergo cadit quæstio].

Resp. [Concedo majorem sed nego minorem.]

Opp. [Probo minorem.] Si eadem ratio quæ prius eum docebat dignoscere tangendo inter cubum et globum eum etiam docebit intuendo recte dignoscere, valent minor et argumentum. [Sed eadem ratio quæ prius eum docebat dignoscere tangendo inter cubum et globum eum etiam docebit intuendo recte dignoscere. Ergo valent minor et argumentum.]

Watson was subsequently followed on the same side by W. Lax of Trinity as second opponent, and Richard Riley of St John's as third opponent; and it would seem from the tripos list of 1785 that Sewell was altogether overmatched by his antagonists.

The following account of some disputations in 1790 is taken from a letter by William Gooch of Caius, who was second wrangler in 1791. It is especially valuable as giving us an undergraduate's view of these exercises. Another letter by him descriptive of the senate-house examination in 1791 is printed in the next chapter. The letter in question is dated Nov. 6, 1790, and after some gossip about himself he goes on

Peacock kept a very capital Act indeed and had a very splendid Honor of which I can't remember a Quarter, however among a great many other things, Lax told him that "Abstruse and difficult as his Questions were, no Argument (however well constructed) could be brought against any Part of them, so as to baffle his inimitable Discerning & keen Penetration" &c. &c. &c.—However the Truth was that he confuted all the Arguments but *one* which was the 1st Opponent's 2nd Argument,—Lax lent him his assistance too, yet still he didn't see it, which I was much surpris'd at as it seem'd easier than the Majority of the rest of the Args—Peacock with the Opponents return'd from the Schools to my Room to tea, when (agreeable to his usual ingenuous Manner) he mention'd his being in the Mud about Wingfield's 2nd argument, & requested Wingfield to read it to him again & *then* upon a little consideration he gave a very ample answer to it.—I was *third* opponent only and came off with "*optime quidem disputasti*" i.e. "you've disputed excellently indeed" (quite as much as is ever given to a third opponency)—I've a first opponency for Novr 11th under Newton against Wingfield & a second opponency for Novr 19th under Lax against Gray of Peter-House. Peacock is Gray's first opponent & Wingfield his third, so master Gray is likely to be pretty well baited. His third Question (of all things in the world) is to defend Berkley's immaterial System.

Mrs Hankinson & Miss Paget of Lynn are now at Cambridge, I drank tea & supp'd with them on Thursday at Mr Smithson's (the Cook's of St. Johns Coll.) & yesterday I din'd drank tea and supp'd there again with the same Party, and to day I'm going to meet them at Dinner at Mr Hall's of Camb. Hankinson of Trin. (as you may suppose) have (*sic*) been there too always when I have been there; as also Smithson of Emmanuel Coll. (son of this Mr Smithson). Miss Smithson is a very accomplished girl, & a great deal of unaffected Modesty connected with as much Delicacy makes her very engaging.—She talks French, and plays well on the Harpsichord. Mrs H. will continue in Camb. but for a day or two longer or I should reckon this a considerable Breach upon my Time;—However I never can settle well to any thing but my Exercises when I have any upon my Hands, and I'm sure I don't know what purpose 'twould answer to fagg *much* at my Opponencies, as I doubt whether I should keep *at all* the better or the worse they being upon subjects I've long been pretty well acquainted with.—Yet I'm resolv'd when I've kept my first Opponency next thursday *if possible* to think nothing of my 2nd (for friday se'nnight) till within a day or two of the time—One good thing is I can now have no more, so I've the luck to be free from the schools betimes, for the term doesn't end till the middle of Decr.[1]

My readers may be interested to know that Gooch was quite captivated by Miss Smithson, and he intended to propose to her on his return from the astronomical expedition sent out by the government in 1791—3, in which he took part. He was captured by the South Sea islanders in May, 1792, and murdered before assistance could reach him.

The following list of subjects of acts known to have been kept between 1772 and 1792 is taken from Wordsworth. Some were chosen more than once. The questions on mathematics were as follows.

Newton's *Principia*, book I, section i; book I, sections ii and iii; book I, section iii; book I, section vii; book I, section viii; book I, section xii, props. 1—5; book I, section xii, props. 39 and 40; book I, section xii, prop. 66 and one or more corollaries. Cotes's *Harmonia mensurarum*, prop. 1. Cotes's theorem on centripetal force. Cotes's proposition on the five trajectories. The path of a projectile is a parabola. Halley's determination of the solar parallax. Correction of the aberration of rays by conic sections. The method of fluxions. Smith de focalibus distantibus. Maclaurin, chapter III, sections 1—8 and 11—22. Morgan on mechanical forces. Morgan on the inclined plane. Hamilton on vapour.

The questions on philosophy were as follows.

Berkeley on sight and touch. Montesquieu Laws, chapter I, section i. Locke on faith and reason. Can matter think? The signification of words. Wollaston on happiness. From Paley, On penalties; On happiness; On promises. Free press. Imprisonment for debt. Duelling. The slave trade. Common ideas do not enter by different senses. Composite ideas have no absolute existence. The immortality of the soul may be inferred by the light of nature. The immortality of the soul may be inferred by the light of nature, but no more than that of other animals. The soul is immaterial. Omnia nostrâ de causâ facimus.

A candidate was not however allowed to offer any question. Thus a proposition taken out of Euclid's *Elements* was gene-

[1] *Scholae academicae*, 321—22.

rally rejected by the moderators, probably because of the difficulty of arguing against its correctness. In 1818 as a great concession a questionist was allowed to "keep" in the eleventh book of Euclid. The moderators also refused to allow the maintenance of any doctrine which they regarded as immoral or heretical. Thus when Paley of Christ's, in 1762, proposed for his theses the subjects that punishment in hell did not last throughout eternity, and that a judicial sentence of death for any crime was unjustifiable they were rejected; whereupon he upheld the opposite views in the schools, leaving to his opponents the duty of sustaining his original propositions.

Of the disputations in 1819 Whewell, who was then moderator, writes as follows. "They are held between undergraduates in pulpits on opposite sides of the room, in Latin and in a syllogistic form. As we are no longer here in the way either of talking Latin habitually or of reading logic, neither the one nor the other is very scientifically exhibited. The syllogisms are such as would make Aristotle stare, and the Latin would make every classical hair in your head stand on end. Still it is an exercise well adapted to try the clearness and soundness of the mathematical ideas of the men, though they are of course embarrassed by talking in an unknown tongue....It does not, at least immediately, produce any effect on a man's place in the tripos, and is therefore considerably less attended to than used to be the case, and in most years is not very interesting after the five or six best men[1]."

Even to the last they sometimes led to a brilliant passage of arms. Thus Richard Shilleto of Trinity College (B.A. 1832, and subsequently a fellow of Peterhouse), kept an act on the well-worn subject as to whether suicide was justifiable[2]. *Quid est suicidium*, said he, *ut Latine nos loquamur nisi suum caesio?* and then he went on to defend it on the

[1] See vol. II. pp. 35, 36 of Todhunter's *Life of Whewell*, London, 1876.

[2] The story is told differently by Wordsworth, but I give it as I have heard it. Suicidium was the scholastic translation of suicide.

ground that roast pig and boiled ham were delicacies appreciated by all. His opponent, a Johnian and good mathematician but ignorant of classics, could not understand a word of this, but the moderator, Francis Martin of Trinity, entered into the spirit of the fun and himself carried on the discussion. In earlier times (and even a few years previously) the acts were a serious matter, and a joke such as this would not have been tolerated.

The form in which they were carried out required a knowledge of formal logic, and (at least) a smattering of conversational Latin; and till within a few years of their abolition in 1839, the publicity of the discussion ensured the most thorough preparation. This previous preparation was the more necessary as the respondent had to answer off-hand any objection from any source, or any apparent argument however fallacious, which the opponent (in general previously prompted by his tutor) might bring against his thesis.

Thus De Morgan writing about his act kept in 1826 says, "I was badgered for two hours with arguments given and answered in Latin,—or what we call Latin—against Newton's first section, Lagrange's derived functions, and Locke on innate principles. And though I took off everything, and was pronounced by the moderator to have disputed *magno honore*, I never had such a strain of thought in my life. For the inferior opponents were made as sharp as their betters by their tutors, who kept lists of queer objections drawn from all quarters[1]." James Devereux Hustler, the third wrangler of 1806 and subsequently a tutor of Trinity, had a special reputation for prompting men with such objections (see p. 113).

I believe that so long as the discussion was a real one and carried on in the language of formal logic (which prevented the argument wandering from the point), it was an admirable training, though to be productive of the best effects it required a skilled moderator. It not only gave considerable dialectical

[1] See p. 305 of the *Budget of paradoxes* by A. De Morgan, London, 1872.

practice but was a corrective to the thorough but somewhat narrow training of the tripos.

Had the language of the discussions been changed to English, as was repeatedly urged from 1774 onwards, these exercises might have been kept with great advantage, but the barbarous Latin and the syllogistic form in which they were carried on prejudiced their retention. I do not know whether disputations are now used in any university, except as a more or less formal ceremony, after a man's ability has been tested in other ways; but I am told that they still form a part of the training in some of the Jesuit colleges where the students have to maintain heresies against the professors, and that the directors of those institutions have a high opinion of their value.

About 1830 a custom grew up for the respondent and opponents to meet previously and arrange their arguments together. The whole ceremony then became an elaborate farce and was a mere public performance of what had been already rehearsed. Accordingly the moderators of 1840, T. Gaskin and T. Bowstead, took the responsibility of discontinuing them. Their action was singularly high-handed, as a report of May 30, 1838, had recommended that the moderators should continue to be guided by these exercises.

No one, however distinguished, appeared more than twice as a respondent and twice in each grade of opponency, that is, eight times altogether—some of the exercises being performed in the Lent and Easter terms of the third year of residence, and the remainder in the October term of the fourth year. The non-reading men were perhaps only summoned once or twice, and before 1790 fellow-commoners[1] seemed to have been excused all attendance.

[1] The earliest certain instance of a fellow-commoner presenting himself for the senate-house examination is that of T. Gisborne of St John's, who was sixth wrangler in 1780. The first known case of a fellow-commoner appearing in the schools is that of James Scarlett (Lord Abinger) of Trinity, who took a poll B.A. degree in 1790. Before that time their

By the Elizabethan code every student before being admitted to a degree had to swear that he had performed all the statutable exercises. The additional number thus required to be performed were kept by what was called *huddling*. To do this a regent took the moderator's seat, one candidate then occupied the respondent's rostrum, and another the opponent's. *Recte statuit Newtonus*, said the respondent. *Recte non statuit Newtonus*, replied the opponent. This was a disputation, and it was repeated a sufficient number of times to count for as many disputations. The men then changed places, and the same process was repeated, each maintaining the contrary of his first assertion—an admirable practice, as De Morgan observed, for those who were going to enter political life. Jebb[1] asserts that in his time (1772) a candidate in this way could as a respondent read two theses, propound six questions, and answer sixteen arguments against them, all in five minutes.

Throughout the eighteenth century the ceremony of entering the questions (see pp. 147, 155) was purely formal. So also were the quadragesimal exercises, which it will be remembered were held after Ash-Wednesday, and therefore after the degree of bachelor had been conferred. All of these were huddled. The proctor generally asked some question such as *Quid est nomen?* to which the answer usually expected was *Nescio*. In these exercises more license was allowable, and if the proctor could think of any remark which he was pleased to consider witty, particularly if there was any play on words in it, he was at liberty to give free scope to his fancy. Some of the repartees to which these personal remarks gave rise have been preserved. For example, J. Brasse, of Trinity, who was sixth wrangler in 1811, was accosted with the question, *Quid est æs?* to which he answered, *Nescio nisi finis examinationis*.

appearance was optional, but Thomas Jones of Trinity, the senior wrangler of 1779, when moderator in 1786—7, introduced a grace by which fellow-commoners were subjected to the same exercises as other students.

[1] Jebb's *Works*, vol. II. p. 298.

So again Joshua King of Queens' was asked *Quid est rex?* to which he promptly replied, *Socius reginalis*, as ultimately turned out to be the case.

A diligent reader of the literature connected with the university of the eighteenth century may find numbers of these mock disputations; but I will content myself with one more specimen. *Domine respondens*, says the moderator, *quid fecisti in academia triennium commorans? Anne circulum quadrasti?* To which the student shewing his cap with the board broken and the top as much like a circle as anything else, replied: *Minime domine eruditissime: sed quadratum omnino circulavi.*

It should be added that retorts such as these were only allowed in the pretence exercises, and a candidate who in the actual examination was asked to give a definition of happiness and replied an exemption from Payne—that being the name of the moderator then presiding—was plucked "for want of discrimination in time and place."

In earlier times even the farce of huddling seems to have been unnecessary, for the Heads reported to a royal commission in 1675 that it was not uncommon for the proctors to take "cautions for the performance of the statutable exercises, and accept the forfeit of the money so deposited in lieu of their performance."

The exercises for the higher degrees (if kept at all) were universally performed by huddling. The statutable exercises for the M.A. degree were three respondencies, each against a master as opponent, two respondencies against bachelor opponents, and one declamation. In the eighteenth century these had become reduced to a mere form and were all huddled. The usual procedure was to "declaim" two lines of the Æneid or of Virgil's first Eclogue; and then to keep three acts with the formula, *Recte statuit Newtonus, Woodius, et Paleius.* To this the opponent replied (thus keeping three opponencies), *Si non recte statuerunt Newtonus, Woodius, et Paleius cadunt quaestiones: sed non recte statuerunt Newtonus, Woodius, et Paleius: ergo cadunt quaestiones.*

At some time early in the present century (I suspect about 1820) the practice of huddling, at any rate for the master's degree, almost ceased. It was generally felt that it was better to openly violate an antiquated statute than to keep the letter and not the spirit of it. This was largely due to Farish and Peacock.

I may here add that though the standards of education and examination for the bachelor's degree at Oxford during the seventeenth and eighteenth centuries were very far below those at Cambridge, yet the performance of certain exercises for the master's degree was always there enforced, and these to some extent counteracted the evil effects of the absence of any honour examination and of any real disputations for those who took the bachelor's degree.

CHAPTER X.

THE MATHEMATICAL TRIPOS[1].

I TRACED in chapter V. the steps by which mathematics became in the eighteenth century the dominant study in the university. I purpose in this chapter to give a sketch of the rise of the mathematical tripos, that is, of the instrument by which the proficiency of students in mathematics came ultimately to be tested.

The proctors had from the earliest time had the power of questioning a candidate when a disputation was closed. I believe that it was about 1725 that the moderators began the custom of regularly summoning those candidates in regard to whose abilities and position some doubt was felt. In earlier times each candidate had been examined when his act was finished, but now all the candidates to be questioned were present at the same time, and this enabled the moderators to compare one man with another.

An additional reason why it was then desirable to use this latent power was the fact that at that time it had become impossible to get rooms in which all the statutable exercises

[1] The substance of this chapter is taken from my *Origin and history of the mathematical tripos*, Cambridge, 1880. The history of the tripos is also treated in *Of a liberal education*, by W. Whewell, Cambridge, 1848, and in the *Scholae academicae* by C. Wordsworth, Cambridge, 1877. In 1888 Dr Glaisher chose the subject for his inaugural address to the London Mathematical Society: all the more important facts are there brought together in a convenient form, and in some places in the latter part of the chapter I have utilized his summary of the later regulations for the conduct of the examination.

could be properly performed, and many, even of the best men, had no opportunity to shew their dialectical skill by means of the exercises in the schools. This arose from the fact that when George I. in 1710 presented the university with thirty thousand[1] books and manuscripts, there was no suitable place in which they could be arranged. It was accordingly decided to build a new senate-house, and use the old one as part of the library, and meanwhile the books were stored in the schools and the old senate-house. The new building was more than twenty years in course of construction, and during that interval the authorities found it impossible to compel the performance of all the exercises required from candidates for degrees.

During the confusion so caused, the discipline and studies of the university suffered seriously. The new senate-house was opened in 1730, and Matthias Mawson, the master of Corpus, who was vice-chancellor in 1730 and 1731, made a determined effort to restore order. It was however found almost impossible to enforce all the statutable exercises, and there was the less necessity as the examination, which had begun to grow up, supplied a practical means of testing the abilities of the candidates. The advantages of the latter system were so patent that within ten or twelve years it had become systematized into an organized test to which all questionists were liable, although it was still regarded as only supplementary to the exercises in the schools. From the beginning it was conducted in English[2], and accurate lists were made of the order of merit of the candidates; two advantages to which I think its final and definite establishment must be largely attributed.

I therefore place the origin of the senate-house examination about the year 1725; but there are no materials for

[1] The library had been shamefully neglected. It contained at that time less than fifteen thousand volumes: many thousands having been lost or stolen in the two preceding centuries.

[2] I have no doubt that this was the case; but Jebb's statement (made in 1772), if taken by itself, rather implies the contrary.

forming an accurate opinion as to how it was then conducted. It is however probable that for about twenty years or so after its commencement it was looked upon as a tentative and unauthorized experiment. Two changes which were then made caused greater attention to be paid to the order of the tripos list, and thus served to give it more prominence. In the first place, from 1747 onwards the final lists were printed and distributed; from that time also the names of the honorary or proctor's optimes (see p. 170) were specially marked, and it was thus possible, by erasing them, to obtain the correct order of the other candidates. The lists published in the calendars begin therefore with that date, and in the issues for all years subsequent to 1799 the names of those who received these honorary degrees have been omitted. In the second place, it was found possible by means of the new examination to differentiate the better men more accurately than before; and accordingly, in 1753, the first class was subdivided into two, called respectively wranglers and senior optimes, a division which is still maintained.

From 1750 onwards the examination was definitely recognized by the university, and we have now more materials to enable us to judge how it was conducted. It would seem from these that it was presided over by the proctors and moderators, who took all the men from each college together as a class, and passed questions down till they were answered; but it still remained entirely oral, and technically was regarded as subsidiary to the discussions in the schools. As each class thus contained men of very different abilities, a custom grew up by which every candidate was liable to be taken aside to be questioned by any M.A. who wished to do so, and this was regarded as the more important part of the examination. The subjects were mathematics and a smattering of philosophy. At first the examination lasted only one day, but at the end of this period it continued for two days and a half. At the conclusion of the second day the moderators received the reports of those masters of arts who had voluntarily taken part in the examination, and provisionally settled the final list; while the last half-day was used in revising and rearranging the order of merit. In 1763 it was decided that the position of Paley of Christ's as senior in the tripos list to Frere of Caius was to be decided by the senate-house examination and not by the disputations.

During the following years, that is from 1763 to 1779, the traditionary rules which had previously guided the examiners in each year took definite shape, and the senate-house examination and not the disputations became the recognized test by which a man's final place in the list was determined. This was chiefly due to the fact that henceforth the examiners used the disputations only as a means of classifying the men roughly. On the result of their 'acts' (and probably partly also of their general reputation) the candidates were divided into eight classes, each being arranged in alphabetical order. Their subsequent position in the class was determined solely by the senate-house examination. The first two classes comprised all who were expected to be wranglers, the next four classes included the other candidates for honours, and the last two classes consisted of poll men only. Practically any one placed in either of the first two classes was allowed, if he wished, to take an ægrotat senior optime, and thus escape all further examination: this was called *gulphing* it. All the men from one college were no longer taken together, but each class was examined separately and *vivâ voce*. As henceforth all the students comprised in each class were of about equal attainments, it was possible to make the examination more efficient.

A full description of the senate-house examination as it existed in 1772 is extant[1]. It was written by John Jebb, who had been second wrangler in 1757. From this account we find that it had then become usual for the junior moderator of the year and the senior moderator of the preceding year to take the first two or three classes together by themselves at

[1] It is reprinted in §§ 192—204 of Whewell's *Of a liberal education*, second edition, London, 1850.

one table. In a similar way the next four or three classes sat at another table, presided over by the senior moderator of that year and the junior moderator of the preceding one; while the last two classes containing the poll men were examined by themselves. Thus, in all, three distinct sets of papers were set. It is probable that before the examination in the senate-house began a candidate, if manifestly placed in too low a class, was allowed the privilege of challenging the class to which he was assigned. Perhaps this began as a matter of favour, and was only granted in exceptional cases, but a few years later it became a right which every candidate could exercise; and I think that it is partly to its development that the ultimate predominance of the tripos over all the other exercises for degrees is due.

The examination took place in January and lasted three days. The range of subjects for the first or highest class is described by Jebb as follows.

The moderator generally begins with proposing some questions from the six books of Euclid, plane trigonometry, and the first rules of algebra. If any person fails in an answer, the question goes to the next. From the elements of mathematics, a transition is made to the four branches of philosophy, viz. mechanics, hydrostatics, apparent astronomy, and optics, as explained in the works of Maclaurin, Cotes, Helsham, Hamilton, Rutherforth, Keill, Long, Ferguson, and Smith. If the moderator finds the set of questionists, under examination, capable of answering him, he proceeds to the eleventh and twelfth books of Euclid, conic sections, spherical trigonometry, the higher parts of algebra, and Sir Isaac Newton's *Principia*; more particularly those sections which treat of the motion of bodies in eccentric and revolving orbits; the mutual action of spheres, composed of particles attracting each other according to various laws; the theory of pulses, propagated through elastic mediums; and the stupendous fabric of the world. Having closed the philosophical examination, he sometimes asks a few questions in Locke's *Essay on the human understanding*, Butler's *Analogy*, or Clarke's *Attributes*. But as the highest academical distinctions are invariably given to the best proficients in mathematics and natural philosophy, a very superficial knowledge in morality and metaphysics will suffice.

When the division under examination is one of the higher classes, problems are also proposed, with which the student retires to a distant part of the senate-house, and returns, with his solution upon paper, to the moderator, who, at his leisure, compares it with the solutions of other students, to whom the same problems have been proposed.

The extraction of roots, the arithmetic of surds, the invention of divisors, the resolution of quadratic, cubic, and biquadratic equations; together with the doctrine of fluxions, and its application to the solution of questions 'de maximis et minimis,' to the finding of areas, to the rectification of curves, the investigation of the centers of gravity and oscillation, and to the circumstances of bodies, agitated, according to various laws, by centripetal forces, as unfolded, and exemplified, in the fluxional treatises of Lyons, Saunderson, Simpson, Emerson, Maclaurin, and Newton, generally form the subject-matter of these problems.

As the questionists in each class were examined in divisions of six or eight at a time, a considerable number were disengaged at any particular hour. Any master or doctor could then call a man aside and examine him. This separate examination or scrutiny was the test by which the best men were differentiated. Any one who thus voluntarily took part in the examination had to report his impressions to the proper officers. This right of examination was a survival of the part taken by every regent in the exercises of the university; but it constantly gave rise to accusations of partiality[1].

Although the examination lasted but a few days it must have been a severe physical trial to any one who was delicate. It was held in winter and in the senate-house. That building was then noted for its draughts and was not warmed in any way; and we are told that upon one occasion the candidates on entering in the morning found the ink frozen at their desks. The duration of the examination must have been even more trying than the circumstances under which it was conducted. The hours on Monday and Tuesday were from 8 to 9, 9.30 to 11, 1 to 3, 3.30 to 5, and 7 to 9. The evening paper was set in the rooms of the moderator, and wine or tea was provided. The examination on Wednesday ended at 11. On Thursday morning at eight a first list was published with all candidates

[1] See for example Gooch's letter reprinted later on p. 196: see also Bligh's pamphlets of 1780 and 1781.

of about equal merit bracketed, and that day was devoted to arranging the men whose names appeared in the same bracket in their proper order. A man rarely rose above or sunk below his bracket, but during the first hour he had the right, if dissatisfied with his position, to challenge any one above him to a fresh examination in order to see which was the better. At nine a second list came out, and a candidate's power of challenging was then confined to the bracket immediately above his own. Fresh lists revised and corrected came out at 11 a.m., 3 p.m., and 5 p.m. The final list was then prepared. The name of the senior wrangler was announced at midnight, and the rest of the list the next morning. The publication of the list was attended with great excitement.

About this time, circ. 1772, it began to be the custom to dictate some or all of the questions and to require answers to be written. Only one question was dictated at a time, and a fresh one was not given out until some student had solved that previously read—a custom which by causing perpetual interruptions to take down new questions must have proved very harassing. We are perhaps apt to think that an examination conducted by written papers is so natural that the custom is of long continuance. But I can find no record of any (in Europe) earlier than those introduced by Bentley at Trinity in 1702 (see p. 81): though in them it will be observed that every candidate had a different set of questions to answer, so that a strict comparison must have been very difficult. The questions for the Smith's prizes continued until 1830 to be dictated in the manner described above. Even at the present time it is usual to dictate the mathematical papers for the baccalaureate degree in the university of France, but all the questions are read out at once.

In 1779 the senate-house examination was extended to four days, the third day being given up entirely to moral philosophy; at the same time the number of examiners was increased, and the system of brackets recognized as a formal part of the procedure. The right of any M.A. to take part in it, though continuing to exist, was much more sparingly exercised, and I believe was not insisted on after 1785. A candidate who was dissatisfied with the class in which he had been placed as the result of his disputations was henceforth allowed to challenge it before the examination began. This power seems to have been used but rarely; it was however a recognition of the fact that a place in the tripos list was to be determined by the senate-house examination alone, and the examiners soon acquired the habit of settling the preliminary classes without much reference to the previous disputations.

In cases of equality the acts were still taken into account in settling the tripos order; and in 1786 when the second, third, and fourth wranglers came out equal in the examination a memorandum was published that the second place was given to that candidate who *in dialectis magis est versatus*, and the third place to that one who *in scholis sophistarum melius disputavit.*

In 1786 a question set to the expectant wranglers which required the extraction of the square root of a number to three places of decimals is said[1] to have been considered unreasonably hard.

The only papers of this date which as far as I know are now extant are one of the problem papers set in 1785 and one of those set in 1786. These were composed by William Hodson, of Trinity (seventh wrangler in 1764, and vice-master of the college from 1789 to 1793), who was then proctor. The autograph copies from which he gave out the questions were luckily preserved, and have recently been placed in the library of Trinity[2]. They must be almost the last problem papers which were dictated, instead of being printed and given as a whole to the candidates.

[1] See Gunning's *Reminiscences*, vol. I. chap. III. Note however that the *Reminiscences* were not written till 60 or 70 years later; and this statement only represents the author's recollections of the rumours of the time. There are reasons for thinking that the statement is exaggerated.

[2] The Challis Manuscripts, III. 61.

The paper for 1785 is headed by a memorandum to warn candidates to write distinctly and to observe that "at least as much will depend upon the clearness and precision of the answers as upon the quantity of them." The questions are as follows.

1. To prove how many regular Solids there are, what are those Solids called, and why there are no more.

2. To prove the Asymptotes of an Hyperbola always external to the Curve.

3. Suppose a body thrown from an Eminence upon the Earth, what must be the Velocity of Projection, to make it become a secondary planet to the Earth?

4. To prove in all the conic sections generally that the force tending to the focus varies inversely as the square of the Distance.

5. Supposing the periodical times in different Ellipses round the same center of force, to vary in the sesquiplicate ratio of the mean distances, to prove the forces in those mean distances to be inversely as the square of the distance.

6. What is the relation between the 3rd and 7th Sections of Newton, and how are the principles of the 3rd applied to the 7th?

7. To reduce the biquadratic equation $x^4+qx^2+rx+s=0$ to a cubic one.

8. To find the fluent of $\dot{x} \times \sqrt{a^2-x^2}$.

9. To find a number from which if you take its square, there shall remain the greatest difference possible.

10. To rectify the arc DB of the circle $DBRS$. [A figure in the margin shews that an arc of any length is meant.]

The problem paper for 1786 is as follows.

1. To determine the velocity with which a Body must be thrown, in a direction parallel to the Horizon, so as to become a secondary planet to the Earth; as also to describe a parabola, and never return.

2. To demonstrate, supposing the force to vary as $\frac{1}{D^2}$, how far a body must fall both within and without the Circle to acquire the Velocity with which a body revolves in a Circle.

3. Suppose a body to be turned (*sic*) upwards with the Velocity with which it revolves in an Ellipse, how high will it ascend? The same is asked supposing it to move in a parabola.

4. Suppose a force varying first as $\frac{1}{D^3}$, secondly in a greater ratio than $\frac{1}{D^2}$ but less than $\frac{1}{D^3}$, and thirdly in a less ratio than $\frac{1}{D^2}$, in each

of these Cases to determine whether at all, and where the body parting from the higher Apsid will come to the lower.

5. To determine in what situation of the moon's Apsids they go most forwards, and in what situation of her Nodes the Nodes go most backwards, and why?

6. In the cubic equation $x^3+qx+r=0$ which wants the second term; supposing $x=a+b$ and $3ab=-q$, to determine the value of x.

7. To find the fluxion of $x^r \times (y^n+z^m)^{\frac{1}{q}}$.

8. To find the fluent of $\dfrac{a\dot{x}}{a+x}$.

9. To find the fluxion of the mth power of the Logarithm of x.

10. Of right-angled Triangles containing a given Area to find that whereof the sum of the two legs $AB+BC$ shall be the least possible. [This and the two following questions are illustrated by diagrams. The angle at B is the right angle.]

11. To find the Surface of the Cone ABC. [The cone is a right one on a circular base.]

12. To rectify the arc DB of the semicircle DBV.

I insert here the following letter from William Gooch, of Caius, in which he describes his examination in the senate-house in 1791. It must be remembered that it is the letter of an undergraduate addressed to his father and mother, and was not intended either for preservation or publication—a fact which certainly does not detract from its value. His account of his acts in 1790 was printed in the last chapter. This letter is dated January, 1791, and is written almost like a diary.

'*Monday* ½ aft. 12.

We have been examin'd this Morning in pure Mathematics & I've hitherto kept just about even with Peacock which is much more than I expected. We are going at 1 o'clock to be examin'd till 3 in Philosophy.

From 1 till 7 I did more than Peacock; But who did most at Moderator's Rooms this Evening from 7 till 9, I don't know yet;—but I did above three times as much as the Senr Wrangler last year, yet I'm afraid not so much as Peacock.

Between One & three o'Clock I wrote up 9 sheets of Scribbling Paper so you may suppose I was pretty fully employ'd.

Tuesday Night.

I've been shamefully us'd by Lax to-day;—Tho' his anxiety for Peacock must (of course) be very great, I never suspected that his Partially (*sic*) w^d get the better of his Justice. I had entertain'd too high an opinion of him to suppose it.—he gave Peacock a long private Examination & then came to me (I hop'd) on the same subject, but 'twas only to *Bully* me as much as he could,—whatever I said (tho' right) he tried to convert into Nonsense by seeming to misunderstand me. However I don't entirely dispair of being first, tho' you see Lax seems determin'd that I shall not.—I had no Idea (before I went into the Senate-House) of being able to contend at all with Peacock.

Wednesday evening.

Peacock & I are still in perfect Equilibrio & the Examiners themselves can give no guess yet who is likely to be first;—a New Examiner (Wood of St. John's, who is reckon'd the first Mathematician in the University, for Waring doesn't reside) was call'd solely to examine Peacock & me only.—but by this new Plan nothing is yet determin'd.—So Wood is to examine us again to-morrow morning.

Thursday evening.

Peacock is declar'd first & I second,—Smith of this Coll. is either 8th or 9th & Lucas is either 10th or 11th.—Poor Quiz Carver is one of the *οἱ πολλοι*;—I'm perfectly *satisfied* that the Senior Wranglership is Peacock's due, but *certainly* not so very indisputably as Lax pleases to represent it —I understand that *he* asserts 'twas 5 to 4 in Peacock's favor. Now Peacock & I have explain'd to each other how we went on, & can *prove indisputably* that it wasn't 20 to 19 in his favor;—I *cannot* therefore be displeas'd for being plac'd second, tho' I'm provov'd (sic) with Lax for his false report (so much beneath the Character of a Gentleman.)—

N.B. it is my very *particular Request* that you don't mention Lax's behaviour to me to any one[1].'

It was about this time that the custom of printing the problem (but not the other) papers was introduced.

Such was the form ultimately taken by the senate-house examination, a form which it substantially retained without alteration for nearly half a century, and which may fairly be considered as the archetype of the numerous competitive examinations now existing in England. It soon became the sole test by which candidates were judged. In 1790 James Blackburn of Trinity, a questionist of exceptional abilities, was informed that in spite of his good disputations he would not be allowed a degree unless he also satisfied the examiners

[1] *Scholae academicae*, 322—23.

in the tripos. He accordingly solved one 'very hard problem,' though in consequence of a dispute with the authorities he refused to attempt any more. In 1799 a further step in the same direction was taken, and it was determined to require from every candidate a knowledge of the first book of Euclid, arithmetic (to fractions), elementary algebra, Locke's *Essay*, and Paley's *Evidences*. A knowledge of the first two books of Euclid, algebra to simple and quadratic equations, and the early chapters of Paley's *Evidences of Christianity* was still considered sufficient to secure a position in the senior optimes.

Since 1796 a calendar containing an account of the university constitution and customs has been annually published. The following garrulous account of the examination in 1802 is taken from the calendar of that year.

On the Monday morning, a little before eight o'clock, the students, generally about a hundred, enter the senate-house, preceded by a master of arts, who on this occasion is styled the father of the college to which he belongs. On two pillars at the entrance of the senate-house are hung the classes and a paper denoting the hours of examination of those who are thought most competent to contend for honours. Immediately after the university clock has struck eight, the names are called over, and the absentees, being marked, are subject to certain fines. The classes to be examined are called out, and proceed to their appointed tables, where they find pens, ink, and paper provided in great abundance. In this manner, with the utmost order and regularity, two-thirds of the young men are set to work within less than five minutes after the clock has struck eight. There are three chief tables, at which six examiners preside. At the first, the senior moderator of the present year and the junior moderator of the preceding year. At the second, the junior moderator of the present and the senior moderator of the preceding year. At the third, two moderators of the year previous to the two last, or two examiners appointed by the senate. The two first tables are chiefly allotted to the six first classes; the third, or largest, to the *οἱ πολλοί*.

The young men hear the propositions or questions delivered by the examiners; they instantly apply themselves; demonstrate, prove, work out and write down, fairly and legibly (otherwise their labour is of little avail) the answers required. All is silence; nothing heard save the voice of the examiners; or the gentle request of some one, who may wish a repetition of the enunciation. It requires every person to use the utmost dispatch; for as soon as ever the examiners perceive any one to have finished his paper and subscribed his name to it another question is immediately given. A smattering demonstration will weigh little in the scale of merit; everything must be fully, clearly, and scientifically brought to a true conclusion. And though a person may compose his paper amidst hurry and embarrassment, he ought ever to recollect that his papers are all inspected by the united abilities of six examiners with coolness, impartiality, and circumspection.

The examiners are not seated, but keep moving round the tables, both to judge how matters proceed and to deliver their questions at proper intervals. The examination, which embraces arithmetic, algebra, fluxions, the doctrine of infinitesimals and increments, geometry, trigonometry, mechanics, hydrostatics, optics, and astronomy, in all their various gradations, is varied according to circumstances: no one can anticipate a question, for in the course of five minutes he may be dragged from Euclid to Newton, from the humble arithmetic of Bonnycastle to the abstruse analytics of Waring. While this examination is proceeding at the three tables between the hours of eight and nine, printed problems are delivered to each person of the first and second classes; these he takes with him to any window he pleases, where there are pens, ink, and paper prepared for his operations.

At nine o'clock the papers had to be given up, and half-an-hour was allowed for breakfast. At half-past nine the candidates came back, and were examined in the way described above till eleven, when the senate-house was again cleared. An interval of two hours then took place. At one o'clock all returned to be again examined. At three the senate-house was cleared for half-an-hour, and, on the return of the candidates, the examination was continued till five. At seven in the evening the first four classes went to the senior moderator's rooms to solve problems. They were finally dismissed for the day at nine, after eight hours of examination. The work on Tuesday was similar to that of Monday; Wednesday was partly devoted to logic and moral philosophy. At eight o'clock on Thursday morning the brackets or preliminary classifications in order of merit, each containing the names of the candidates placed alphabetically, were hung upon the pillars. The examination that day was devoted to arranging the men in each bracket in their proper order: but every candidate had the right to challenge any one whose name appeared in the bracket immediately above his own. If he proved himself the equal of the man so challenged his name was transferred to the upper bracket. To challenge and then to fail to substantiate the claim to removal to a higher bracket was considered rather ridiculous. Fresh editions and revisions of the brackets were published at 9 a.m., 11 a.m., 3 p.m., and 5 p.m., according to the results of the examination during that day. At five the whole examination ended. The proctors, moderators, and examiners then retired to a room under the public library to prepare the list of honours, which was sometimes settled without much difficulty in a few hours, but sometimes not before two or three the next morning. The name of the senior wrangler was generally published at midnight.

In 1802, there were eighty-six candidates for honours, and they were divided into fifteen brackets, the first and second brackets containing each one name only, and the third bracket four names.

Until 1883 the tripos papers of the current year were printed in the calendar. The papers from 1801 to 1820 were also published separately under the title *Cambridge problems; being a collection of the printed questions proposed to the candidates...at the general examinations from* 1801 *to* 1820 *inclusive.* As complete sets of all the problems set to each of the classes are now rare, I propose to print here the whole of the problem papers set in 1802.

MONDAY MORNING PROBLEMS.—Mr. PALMER.

First and second classes (i.e. the expectant wranglers).

1. Given the three angles of a plane triangle, and the radius of its inscribed circle, to determine its sides.

2. The specific gravities of two fluids, which will not mix, are to each other as $n : 1$, compare the quantities which must be poured into a

cylindrical tube, whose length is (a) inches, that the pressures on the concave surfaces of the tube, which are in contact with the fluids, may be equal.

3. Determine that point in the arc of a quadrant from which two lines being drawn, one to the centre and the other bisecting the radius, the included angle shall be the greatest possible.

4. Required the linear aperture of a concave spherical reflector of glass, that the brightness of the sun's image may be the same when viewed in the reflector and in a given glass lens of the same radius.

5. Determine the evolute to the logarithmic spiral.

6. Prove that the periodic times in all ellipses about the same center are equal.

7. The distance of a small rectilinear object from the eye being given, compare its apparent magnitude when viewed through a cylindrical body of water with that perceived by the naked eye.

8. Find the fluents of the quantities $\frac{d\dot{x}}{x(a^2-x^2)}$, and $\frac{h\dot{y}}{y(a+y)^{\frac{3}{2}}}$.

9. Through what space must a body fall internally, towards the centre of an ellipse, to acquire the velocity in the curve?

10. Find the principal focus of a globule of water placed in air.

11. Determine, after Newton's manner, the law of the force acting perpendicular to the base, by which a body may describe a common cycloid.

12. Find the area of the curve whose equation is $xy=a^x$.

13. What is the value of q that force $\times$ (period)$^2=q\times$ radius of circle?

14. Two places, A and B, are so situated that when the sun is in the northern tropic it rises an hour sooner at A than at B; and when the sun is in the southern tropic it rises an hour later at A than at B. Required the latitudes of the places.

15. From what point in the periphery of an ellipse may an elastic body be so projected as to return to the same point, after three successive reflections at the curve, having in its course described a parallelogram?

MONDAY AFTERNOON PROBLEMS.—MR. DEALTRY.

Third and fourth classes (i.e. the expectant senior optimes).

1. Inscribe the greatest cylinder in a given sphere.

2. Rays, which pass through a globe at equal distances from the centre, are turned equally out of their course.—Required a proof.

3. Given a declination of the sun and the latitude of the place, to find the duration of twilight.

4. A cylindrical vessel, 16 feet high, empties itself in four hours by a hole in the bottom.—What space does the surface describe in each hour?

5. Prove that if two circles touch each other externally, and parallel diameters be drawn, the straight lines, which join the opposite extremities of these diameters, will pass through the point of contact.

6. A ball, whose elasticity : perfect elasticity :: n : 1, falls from a given height upon a hard plane, and rebounds continually till its whole motion is lost.—Find the space passed over.

7. If a body revolves in any curve, compare the angular velocity of the perpendicular with that of the distance.

8. How far must a body fall externally to acquire the velocity in a circle, the force varying as the distance?

9. Given the right ascensions and declinations of two stars, to find their distance.

10. Find the velocity with which air rushes into an exhausted receiver.

11. Let the roots of the equation $x^3 - px^2 + qx - r = 0$ be a, b, and c, to transform it into another, whose roots are a^2, b^2, c^2.

12. Find the fluent $\frac{z\dot{z}}{1+2az+z^2}$, a being less than 1; and of $\frac{x\dot{x}}{\sqrt{a^2+x^2}}$.

13. Find that point in the ellipse, where the velocity is a geometric mean between the greatest and least velocities, the force varying $\frac{1}{D^2}$.

14. Determine the position of a line drawn from a given point to a given inclined plane, through which the body will fall in the same time as through the given plane.

15. The equation $x^3 - 5x^2 + 8x - 4 = 0$ has two equal roots.—Find them.

16. Find the sum of the cube numbers $1 + 8 + 27 +$ &c. by the differential method; and sum the following series by the method of increments:

$$1 . 2 + 2 . 3 + 3 . 4 + \text{\&c. } n \text{ terms.}$$

$$\frac{1}{1 . 2} + \frac{1}{2 . 3} + \frac{1}{3 . 4} \text{ \&c. } n \text{ terms and ad infinitum.}$$

17. If half of the earth were taken off by the impulse of a comet, what change would be produced in the moon's orbit?

18. Prove that if the eye be placed in the principal focus of a lens, the image of a given object would always appear the same.

19. Find the time of emptying a given paraboloid by a hole made in the vertex.

20. Find the proportion between the centripetal and centrifugal forces in a curve; and apply the expression to the reciprocal spiral.

MONDAY AFTERNOON PROBLEMS.—Mr. DEALTRY.

Fifth and sixth classes (i.e. the expectant junior optimes).

1. Prove that an arithmetic mean is greater than a geometric.

2. Every section of a sphere is a circle.—Required a proof.

3. If $\frac{3}{4}$ of an ell of Holland cost $\frac{1}{4}$£. what will $12\frac{3}{8}$ ells cost?

4. Prove the method of completing the square in a quadratic equation.

5. Take away the second term of the equation $x^2 - 12x + 5 = 0$.

6. Inscribe the greatest rectangle in a given circle.

7. Sum the following series:

$$1 + 3 + 5 + 7 + \text{\&c. to } n \text{ terms.}$$

$$3 - 1 + \frac{1}{3} - \frac{1}{9} + \text{\&c. ad inf.}$$

$$\frac{1}{1.2.3}+\frac{1}{2.3.4}+\frac{1}{3.4.5} \text{ \&c. ad inf.}$$

8. Find the value of x in the following equations:

$$\frac{42x}{x-2}=\frac{35x}{x-3}$$

$$\frac{1}{2}(x+1)+\frac{1}{3}(x+2)=16-\frac{1}{4}(x+3)$$

$$3x^2-14x+15=0.$$

9. In a given circle to inscribe an equilateral triangle.

10. Two equal bodies move at the same instant from the same extremity of the diameter of a circle with equal velocities in opposite semi-circles. Required the path described by the centre of gravity; find the path also when the bodies are unequal.

11. Through what chord of a circle must a body fall to acquire half the velocity gained by falling through the diameter?

12. Given the latitude of the place and the sun's meridian altitude, to find the declination.

13. Given the sun's altitude and azimuth and the latitude of the place, to find the declination and the hour of the day.

14. Prove that the velocity in a parabola : velocity in a circle at the same distance :: $\sqrt{2}:1$.

15. How far must a body fall internally to acquire the velocity in a circle, the force varying $\frac{1}{D^2}$?

MONDAY EVENING PROBLEMS.—Mr. DEALTRY.

First, second, third, and fourth classes.

1. Find four geometric means between 1 and 32, and three arithmetic means between 1 and 11.

2. Suppose a straight lever has some weight, and at one end a weight is suspended equal to that of the lever; where must the fulcrum be placed, that there may be an equilibrium?

3. Determine the latitude of the place, where the sun's meridian altitude is 73°. 24′. 13″, its declination south being 16°. 36′. 47″.

4. If Q represent the length of a quadrant, whose radius is R, and the force vary $\frac{1}{D^2}$, the time of descent half way to the centre of force : the time through the remaining half :: $Q+R:Q-R$. Required a proof.

5. P and W represent two weights hung over a fixed pulley; supposing P to descend, what space will it describe in t'', the inertia of the pulley being taken into the account?

6. If a pendulum, whose length is 40 inches would oscillate in 1″ at the pole of a sphere, the radius of which is 4000 miles; what must be the time of rotation round its axis, that the same pendulum at the equator may oscillate twice in 3″?

7. A given cone is immersed in water with its vertex downward;

what part of the axis will be immersed, if the specific gravity of the fluid : that of the cone :: 8 : 1?

8. The axis of a wheel and axle is placed in a horizontal position, and a weight y, which is applied to the circumference of the axle, is raised by the application of a given moving force p applied to the circumference of the wheel; given the radii of the wheel and axle, it is required to assign the quantity y, when the moment generated in it in a given time is a maximum, the inertia of the wheel and axle not being considered.

9. Would Venus ever appear retrograde according to the Tychonic system?

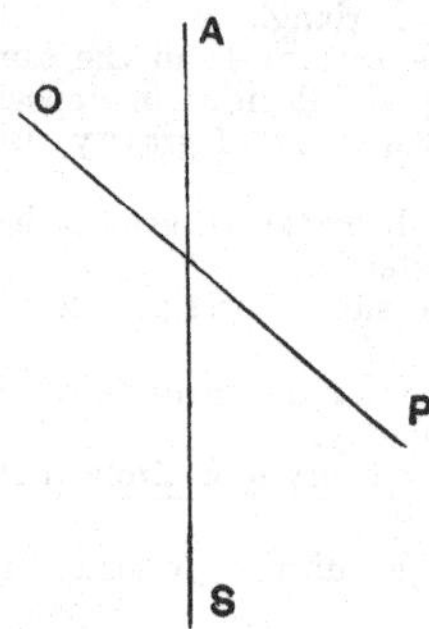

10. A perfectly elastic ball begins to fall from a given distance SA in a right line towards the centre of force S, the force varying $\frac{1}{D^2}$; in its descent, it impinges upon a hard plane OP inclined to SA at a given angle, and after describing a certain curve comes to the plane on the other side, and is then reflected to the center; find the nature of this curve; and determine the whole time of descent to the center S in terms of the periodic time of a body revolving in a circle at the distance SA.

11. Let parallel rays be refracted through two contiguous double convex lenses; find the focal length on the supposition that the radii of all the surfaces are equal, and the sine of incidence : sine of refraction :: 5 : 4.

12. Given the latitude of the place and the declination of the sun, the former being less than the latter; to find at what time of the day the shadow of a stick would be stationary, and how far it would afterwards recede on the horizontal plane.

13. Transform the equation $x^n - px^{n-1} + qx^{n-2} - \&c. = 0$ into one, whose roots are the reciprocals of the sum of every $n-1$ roots of the original equation.

14. A body descends down the cycloidal arc AM, the base AL being parallel to the horizon and M the lowest point of the cycloid; determine that point where its velocity in a direction perpendicular to the horizon is a maximum.

15. Construct the equation $a^2y - x^2y - a^3 = 0$.

16. Compare the time of descent to the center in the logarithmic spiral with the periodic time in a circle, whose radius is equal to the distance from which the body is projected downward.

17. Given the difference of altitudes of two stars, which are upon the meridian at the same time, and their difference of altitudes and difference of azimuths an hour afterwards, to find the latitude of the place.

18. A person's face in a reflecting concave decreases to the principal focus, and then increases in going from it.—Required a demonstration.

19. Prove that the mean quantity of the disturbing force of S upon P, in the 66th proposition of Newton, during one revolution of P round T, is ablatitious, and equal to half the mean addititious force.

20. The time of the sun's rising is the time which elapses between the appulse of the upper and under limb of the sun's disc to the horizon; given the sun's apparent diameter and the latitude of the place, it is required to determine the declination, when this time is a minimum.

21. Through a given point situate between two right lines given in position, to draw a third line cutting them in such a manner, that the rectangle under the parts intercepted between the point and the two lines may be a minimum.

22. Let a spherical body descend in a fluid from rest; having given the diameter of the sphere, and its specific gravity with reference to that of the fluid, it is required to assign the velocity of the sphere at any given point of the space described.

23. The distance of the centre of gravity from the vertex of a solid formed by the revolution of a curved surface is $\frac{4}{5}$ of its axis.—Determine the nature of the generating curve.

24. Suppose a given cylindrical vessel filled with water to revolve with a given angular velocity round its axis.—Required the quantity contained in the cylinder, when the water and cylinder are relatively at rest.

25. Sum the following series:

$$\frac{10}{1.2.3.4}+\frac{14}{2.3.4.5}+\frac{18}{3.4.5.6}+\&c. \text{ to } n \text{ terms and ad inf.}$$

$$\frac{5}{1.2.3}\times\frac{1}{2^2}+\frac{6}{2.3.4}\times\frac{1}{2^3}+\frac{7}{3.4.5}\times\frac{1}{2^4}+\&c. \text{ ad inf.}$$

$$\frac{1}{1.3}-\frac{1}{3.5}+\frac{1}{5.7}-\&c. \text{ ad inf.}$$

26. Given the fluent $(a+cz^n)^m\times z^{pn+n-1}\dot{z}$
to find the fluent $(a+cz^n)^{m+1}\times z^{pn-1}\dot{z}$.

Required also fluent $\dfrac{\dot{x}\sqrt{a^2+x^2}}{x^3}$; and of $\dfrac{z^\theta\dot{z}}{1+mz}$, θ being a whole positive number.

TUESDAY MORNING PROBLEMS.—Mr. DEALTRY.

First and second classes.

1. Inscribe the greatest cone in a given spheroid.

2. A parabolic surface is immersed vertically in a fluid, whose density increases as the depth, with its base contiguous to the surface of the fluid; find upon which of the ordinates to the axis there is the greatest pressure.

3. Solve the equation $x^3-px^2+qx-r=0$, whose roots are in geometric progression.

4. Suppose the reflecting curve to be a circular arc, and the focus of incident rays in the circumference of the circle, to find the nature of the caustic.

5. If the sine of incidence : sine of refraction :: $m:n$, required the focal length of a hemisphere, the rays falling first on the convex side.

6. If the subtangent of a logarithmic curve be equal to the sub-

tangent of the reciprocal spiral, prove that the arc intercepted between any two rays in the spiral is equal to the arc intercepted between any two ordinates of the curve respectively equal to the former.

7. In what direction must a body be projected from the top of a given tower with a given velocity, so that it may fall upon the horizontal plane at the greatest distance possible from the bottom of the tower?

8. Draw an asymptote to the elliptic spiral.

9. If water or any fluid ascends and descends with a reciprocal motion in the legs of a cylindrical canal inclined at any angle, to find the length of a pendulum which will vibrate in the same time with the fluid.

10. Find the fluent $vx\dot{x}$, where $v = \text{hyp. log.}\ (x + \sqrt{x^2 + a^2})$.

11. The centrifugal force at the equator arising from the rotation of the earth round its axis : the centrifugal force in any parallel of latitude :: $(\text{rad.})^2 : (\text{sine.})^2$ of the co-latitude.—Required a proof.

12. Given the latitudes of two places together with their difference of longitudes, to find the declination of the sun, when it sets to the two places at the same time.

13. Required the equation to a curve, whose subtangent is equal to n times its abscissa.

14. If the force vary $\dfrac{1}{D^{n+1}}$, how far must a body fall externally to acquire the velocity in any curve, whose chord of curvature at the point of projection is c? and apply the expression to the parabola and logarithmic spiral.

TUESDAY AFTERNOON PROBLEMS.—Mr. PALMER.

Third and fourth classes.

1. Find the value of £123333, &c. (*sic*)

2. Determine geometrically a mean proportional between the sum and difference of two given straight lines.

3. What is the general form of parallelograms, whose diameters cut each other at right angles?

4. Investigate the area of a circle, whose diameter is unity; and prove that the areas of different circles are in a duplicate ratio of their diameters.

5. Divide a given line into two parts, such that their product multiplied by their difference may be a maximum.

6. Prove that in any curve the velocity : velocity in a circle at the same distance (SP) :: $\sqrt{\text{chord of curvature}} : \sqrt{2SP}$.

7. A body projected from one extremity of the diameter of a circle, at an angle of 45°, strikes a marked place in the center. Required the velocity of projection and greatest altitude.

8. Find the area of a curve whose equation is $y = \dfrac{a^3}{a^2 - x^2}$.

9. In how many years will the interest due upon £100 be equal to the principal, allowing compound interest?

10. Admitting the periods of the different planets to be in a sesqui-

plicate ratio of the principal axes of their orbits, shew that they are attracted towards the sun by forces reciprocally proportional to the squares of their several distances from it.

11. Prove that in the course of the year the sun is as long above the horizon of any place as he is below it.

12. Determine the limits within which an eclipse of the sun or moon may be expected; and shew what is the greatest number of both which can happen in one year.

13. Prove that the time in which any regular vessel will freely empty itself : time in which a body will freely fall down twice its height :: area of base : area of orifice.

14. Find the fluents of $\frac{x\dot{x}}{\sqrt{a-x}}$; $\frac{x^2\dot{x}}{a-x}$.

15. Find the principal focus of a lens; and shew how an object may be placed before a double convex lens, that its image may be inverted and magnified so as to be twice as great as the object.

16. Prove that Cardan's rule fails unless two roots of the proposed cubic be impossible; and determine whether that rule be applicable to the equation $x^3-237x-884=0$.

17. Deduce Newton's general expression in Sect. 9, for the force in the moveable orbit.

18. Define logarithms, and explain their use; also, prove that

$$\log A\times B=\log A+\log B.$$

19. Explain the different kinds of parallax; and shew from the want of parallax in the fixed stars, that their distance from the earth bears no finite ratio to that of the sun.

TUESDAY AFTERNOON PROBLEMS.—Mr. PALMER.

Fifth and sixth classes.

1. How many yards of cloth, worth 3*s.* 7½*d.* per yard, must be given in exchange for 935½ yards, worth 18*s.* 1½*d.* per yard?

2. Find the interest of £873. 15*s.* 0*d.* for 2½ years at 4¾ per cent.

3. Prove that the diameters of a square bisect each other at right angles.

4. Prove the opposite angles of a quadrilateral figure inscribed in a circle equal to two right angles.

5. Prove that if $A \propto B$ when C is given, and $A \propto C$ when B is given, when neither B nor C is given, $A \propto BC$.

6. Prove radius a mean proportional between tangent and cotangent; and that sine × cosine $\propto$ (sine)^2 of twice the angle.

7. Given the sine of an angle, to find the sine of twice that angle.

8. Prove that in the parabola (ordinate)^2 = abscissa × parameter.

9. Extract the square root of a^3-x^3.

10. Solve the equation $3x^2-19x+16=0$.

11. Prove that motion when estimated in a given direction is not increased by resolution.

12. Find the ratio of $P : W$ when every string in a system of pullies is fastened to the weight.

13. Prove that time of oscillation $\propto \dfrac{\sqrt{\text{length}}}{\sqrt{\text{force}}}$.

14. Prove that when a fluid passes through pipes kept constantly full, velocity $\propto$ inversely as area of section.

15. Define the centre of a lens; and find the centre of a meniscus.

16. Find the fluxion of $\sqrt{a^3+x^3-\sqrt{a^2-x^2}}$.

17. Prove elevation of the equator above the horizon = co-latitude.

18. Prove that sagita $\propto$ (arc)^2.

19. Prove that in the same orbit velocity $\propto$ inversely as perp.

TUESDAY EVENING PROBLEMS.—Mr. PALMER.

First, second, third, and fourth classes.

1. When £100 stock may be purchased in the 3 per cents. for £59½, at what rate may the same quantity of stock be purchased in the 5 per cents. with equal advantage?

2. A ball of wood being balanced in air by the same weight of iron, how will the equilibrium be affected when the bodies are weighed in vacuo? and by what weight of wood, properly disposed, may the equilibrium be restored?

3. Investigate the value of the circumference of a circle whose radius is unity.

4. Compare the areas of the parabolas described by two bodies projected together from the same point, and with the same velocity, towards a mark situated in an horizontal plane, the angles of elevation being to each other :: 2 : 1.

5. Prove the rule for finding the quadratic divisors of any equation; and apply it to the equation $x^4-17x^3+88x^2-172x+112=0$.

6. On what point of the compass does the sun rise to those who live under the equinoctial, when he is in the northern tropic?

7. How many equal circles may be placed around another circle of the same diameter, touching each other and the interior circle?

8. Determine the resistance of the medium in which a body by an uniform gravity may describe a parabolic orbit?

9. Prove that a body moving in the reciprocal spiral, approaches or leaves the centre uniformly.

10. Find the velocity and time of flight of a body projected from one extremity of the base of an equilateral triangle, and in the direction of the side adjacent to that extremity towards an object placed in the other extremity of the base.

11. Define similar curves; and prove that conterminous arcs of such curves have their chords of curvature at the point of contact in a given ratio.

12. Compare the time of a revolution about the center of a given ellipse, with that about its focus.

13. Find the attraction of a corpuscle placed in the axis of a cylindrical superficies, whose particles attract in an inverse duplicate ratio of the distance.

14. Prove that if the center of oscillation of a pendulum be made

the point of suspension, the former point of suspension becomes the center of oscillation.

15. Determine the content of the solid generated by a semicircle revolving about a tangent parallel to its base.

16. Find the fluents of $\frac{x^{-2}\dot{x}}{\sqrt{a-x}}$; $y^{\frac{3n}{2}-1}\dot{y}\sqrt{a^n-y^n}$; $\frac{v\dot{v}}{(a+v)(a^2+v^2)}$.

17. Sum the series $1-\frac{1}{2^3}+\frac{1}{2^5}-\frac{1}{2^7}+$&c. ad inf. and also to n terms.

$\frac{1}{1.5}+\frac{1}{1.6}+\frac{1}{3.7}+$&c. to n terms. $\frac{1}{1.3}+\frac{1}{3.7}+\frac{1}{5.11}+$&c. ad inf.

18. Required the sun's place in the ecliptic, when the increment of his declination is equal to that of his right ascension.

19. Prove that the force by which a body may describe a curve, whose ordinates are parallel, is proportioned to $\pm\ddot{y}$; and determine the quantity q such that force $=q\times\pm\ddot{y}$.

20. Compare the times in which a cylinder, whose axis is parallel to the horizon, will discharge the first and last half of its content through an orifice in its lowest section.

21. Prove that the image of a straight line immersed in water appears concave to an eye placed anywhere between the extremities of the line.

22. At what distance from the earth would the apparent brightness of the moon be equal to that of Saturn and his ring together, supposing the apparent brightness of Saturn to that of his ring :: 2 : 1?

No problems were ever set to the seventh and eighth classes, which contained the poll men. None of the book-work papers of this time are now extant, but it is believed that they contained no riders. It will be seen from the above specimens that many of the so-called problems were really pieces of book-work or easy riders: it must however be remembered that the text-books then in circulation were inferior and incomplete as compared with modern ones.

A few minor changes in the senate-house examinations were made in the following years. In 1808 a fifth day was added to the examination. Of the five days thus given up to it, three were devoted to mathematics, one to logic, philosophy, and religion, and one to the arrangement of the brackets. Apart from the evening paper, the examination on each of the first three days lasted six hours. Of these eighteen hours eleven were assigned to book-work and seven to problems. In 1800 the first four classes had been allowed to take the problem papers, and in 1818 they were opened to all the candidates for honours, i.e. the first six classes, and set from 6 to 10 in the evening: the hours of examination being thus extended to ten a day.

Some observations on the tripos examination of 1806 will be found in the letter by Sir Frederick Pollock to which reference has been already made (see p. 112). A letter from Whewell, dated January 19, 1816, describes his examination in the senate-house[1]. It was at this time that the character of the examination was changing and that the differential notation and analysis were being introduced in the place of fluxions and geometry. The remarks of Peacock and others on this subject have been already quoted (see chapter VII.). Whewell was moderator in 1820, and in a letter to his sister dated Jan. 20, 1820, he describes the examination. There is nothing of any historical interest in his account, save that it shews that many of the questions were still dictated. The letter is as follows[2].

> The examination in the senate-house begins to-morrow, and is rather close work while it lasts. We are employed from seven in the morning till five in the evening in giving out questions and receiving written answers to them; and when that is over, we have to read over all the papers which we have received in the course of the day, to determine who have done best, which is a business that in numerous years has often kept the examiners up the half of every night; but this year is not particularly numerous. In addition to all this, the examination is conducted in a building which happens to be a very beautiful one, with a marble floor and a highly ornamented ceiling; and as it is on the model of a Grecian temple, and as temples had no chimneys, and as a stove or a fire of any kind might disfigure the building, we are obliged to take the weather as it happens to be, and when it is cold we have the full benefit of it—which is likely to be the case this year. However, it is only a few days, and we have done with it.

In the decade from 1820 to 1830 a powerful party arose in the university, as in the country, which desired to overhaul all

[1] See p. 20 of Douglas's *Life of Whewell*, London, 1881.
[2] See p. 56 of Douglas's *Life of Whewell*, London, 1881.

existing methods and regulations. Among other changes the Previous Examination, or Little-Go, was established in 1824, for students in their second year; a reform which was urgently needed, as till then the university required nothing from its undergraduate members until they had entered their third year of residence. The power of granting honorary optime degrees, which had already fallen into abeyance, was abolished. At the same time the classical tripos was founded for those who had already taken honours in mathematics, and the plan of the senate-house examination was re-arranged. Henceforth it is known as the mathematical tripos.

From this time onwards the examination was conducted in each year by four examiners, namely, the two moderators and the two examiners, the moderators of one year becoming as a matter of course the examiners of the next. Thus of the four examiners in each year, two had taken part in the examination of the previous year. The continuity of the examination was well kept up by this arrangement; but it had the effect of causing its traditions to be somewhat punctiliously observed, the papers of each year being, as regards the subjects included, exact counterparts of the corresponding papers of the previous year.

By regulations[1] which were confirmed by the senate on November 13, 1827, and came into operation in January 1828, another day was added, so that the examination in mathematics extended over four days, exclusive of the day of arranging the brackets; the number of hours of examination was twenty-three, of which seven were assigned to problems. On the first two days all the candidates had the same questions proposed to them, inclusive of the evening problems, and the examination on those days excluded the higher and more difficult parts of mathematics, in order, in the words of the report, "that the candidates for honours may not be induced

[1] Most of the analysis here given of the regulations of 1827, 1832, and 1848 is taken from Dr Glaisher's inaugural address to the London Mathematical Society in 1888.

to pursue the more abstruse and profound mathematics, to the neglect of more elementary knowledge." Accordingly, only such questions as could be solved without the aid of the differential calculus were set on the first day, and those set on the second day involved only its elementary applications. The classes were reduced to four, determined as before by the exercises in the schools. The regulations of 1827 are especially important because they first prescribed that all the papers should be printed. They are also noticeable as being the last which gave the examiners power to ask *vivâ voce* questions. After recommending that there be not contained in any paper more questions than well-prepared students have generally been found able to answer within the time allowed for the paper, the report proceeds "but if any candidate shall, before the end of the time, have answered all the questions in the paper, the examiners may at their discretion propose additional questions *vivâ voce*."

At the same time as these changes were made (i.e. in 1828) the examination for the poll degree was separated from the tripos and placed in the following week, with different sets of papers and a different schedule of subjects. It was, however, still nominally considered as forming part of the senate-house examination. It is perhaps worthy of remark that this fiction was maintained till 1858, and those who obtained a poll degree were arranged according to merit into four classes, viz., a fourth, fifth, sixth, and seventh, as if in continuation of the junior optimes or third class of the tripos. Till 1850 all members of the university who took the degree of bachelor of arts were expected to pass what we now call the mathematical tripos, but which was then the only examination held for that degree. The year 1828 therefore shews us the examination dividing into two distinct parts. In 1850 the classical tripos was made independent of the mathematical tripos, and thus provided another and separate avenue to a degree. In 1858 the poll-examination was finally separated from the other part of the mathematical tripos, and provided a third way of obtaining the degree. Since then numerous other ways of obtaining the degree have been established, and it is now possible to get it by shewing proficiency in very special or even technical subjects. I may just add in passing that the examination usually termed "the general" is historically the survival of the old senate-house examination for the poll men; and that in 1852 a third examination, at first called "the professors's examinations," and now known as "the specials," was instituted for all poll men to take at the end of their third year.

New regulations concerning the mathematical tripos were confirmed by the senate on April 6, 1832, and took effect in 1833. The commencement of the examination was placed a day earlier, the duration was extended to five days, and the number of hours of examination on each day was fixed at five and a-half. Twenty hours were assigned to book-work, and seven and a-half to problems. The examination on the first day was confined to subjects that did not require the differential calculus, and only the simplest applications of the calculus were permitted on the second and third days. During the first four days of the examination the same papers were set to all the candidates alike, but on the fifth day the examination was conducted according to classes. No reference was made to *vivâ voce* questions, and the preliminary classification of the brackets only survived in a permission to use it if it were found necessary.

The tripos of 1836 is said to have been the earliest one in which all the papers were marked[1]. In previous years the examiners had partly relied on their impression of the answers given.

The regulations of 1832 were superseded by a new system, which passed the senate on June 2, 1838, and came into operation in January 1839. By these new rules the examination lasted for a week. It began on the Wednesday week

[1] This comes to me on the authority of the late Samuel Earnshaw, the senior moderator of that year.

preceding the first Monday in the Lent term, and ended on the following Tuesday night; and continued every day from nine to half-past eleven in the morning, and from one to four in the afternoon. The list was published on the Friday week following. Of the thirty-three hours of examination, eight and a-half were assigned to problems. Throughout the whole examination the same papers were set to all the candidates. The permissive rule relating to the re-examination of the candidates (a relic of the brackets) was retained in these regulations in the same form as in those of 1832. The examination was for the future confined to mathematics, and "religion" and "philosophy" henceforth disappear from the schedule of subjects. The former of these was, it is true, temporarily reintroduced in 1846 in the form of papers on the New Testament, Paley, and Ecclesiastical history, but as in settling the final list no account was taken of the marks obtained in these papers they were generally neglected. They were accordingly again struck out by a grace of the senate in 1855, and have never been reinstated.

These regulations contain no allusion to the classes, and it was no doubt in accordance with the spirit of these changes that the acts in the schools should be abolished, but they seem to have been discontinued by the moderators of 1839 on their own authority (see p. 183).

A few years later the scheme of the examination was again reconstructed by regulations which came into effect in 1848. The examination, as thus constituted, underwent no further alteration till 1873, and the first three days remain practically unchanged at the present time. The duration of the examination was extended from six to eight days, the first three days being assigned to the elementary and the last five to the higher parts of mathematics. After the first three days there was an interval of a few days at the end of which the moderators and examiners issued a list of those who had so acquitted themselves as to deserve mathematical honours. Only those whose names were contained in this list were admitted to the last five days of the examination. After the conclusion of the examination the moderators and examiners, taking into account the whole eight days, brought out the list arranged in order of merit. No provision was made for any re-arrangement of this list corresponding to the examination of the brackets, which, though forming part of the previous scheme, had been discontinued for some time. An important part of the new regulations was the limitation, by a schedule, of the subjects of examination in the first three days, and of the manner in which the questions were to be answered; the methods of analytical geometry and differential calculus being excluded. In all the subjects contained in this schedule examples and questions arising directly out of the propositions were to be introduced into the papers, in addition to the propositions themselves. Taking the whole eight days, the examination lasted forty-four and a half hours, twelve hours of which were devoted to problems.

In the same year as these regulations came into force, the Board of mathematical studies (consisting of the mathematical professors, and the moderators and examiners for the current and two preceding years) was constituted by the senate. In May 1849 they issued a report in which, after giving a short review of the past and existing state of mathematical studies in the university, they recommended that, considering the great number of subjects occupying the attention of the candidates, and the doubt existing as to the range of subjects from which questions might be proposed, the mathematical theories of electricity, magnetism, and heat should not be admitted as subjects of examination. In the following year they issued a second report, in which they recommended the omission of elliptic integrals, Laplace's coefficients, capillary attraction, and the figure of the earth considered as heterogeneous, as well as a definite limitation of the questions in lunar and planetary theory. In making these recommendations, the Board stated that they were only giving expression to what had become the practice in the examination, and were merely putting before the candidates such results as might have been deduced by any one from a study of the senate-house papers of the preceding years. The Board also recommended that the papers containing book-work and riders should be shortened.

From that time forward their minutes supply a permanent record of the changes gradually introduced into the tripos. Those changes lie beyond the limits of this book.

I may just, in passing, mention a curious attempt which was made in 1854 to assist candidates in judging of the relative difficulty of the questions asked, by informing them of the marks assigned to each question. The marks for the book-work and rider of each question were printed on a little slip of paper which was given to the candidates at the same time as the examination paper[1].

It is not unusual to hear the remark that the scheme of the tripos from 1839 to 1873 was framed so as to discourage those who wished to apply mathematics to physical questions; but that opinion is, I think, framed on a misunderstanding. The university insisted that her mathematical graduates should have a thorough knowledge of all the elementary subjects, and left to them the particular sciences to which they might (if they felt inclined) apply it. It only needs a glance at the tripos lists to see that this course was in no way prejudicial to any branch of mathematical science. Indeed I believe that if the senate had not been so anxious to define exactly what might and what might not be asked, but had allowed the subjects of the examination to grow by the gradual introduction of questions from the more recent applications of mathematics, there is no reason why the regulations of 1841 or of 1848 should not meet all the requirements of the present time. Under those regulations the Cambridge graduate who devoted himself to mathematical research possessed a great advantage

[1] I mention the fact rather because these things are rapidly forgotten than because it is of any intrinsic value. I possess a complete set of slips which came to me from Dr Todhunter.

over his continental colleagues in the wider range of his general mathematical knowledge. That advantage has recently been abandoned, but on the other hand a man on taking his degree is now a specialist in some small part of one branch of the subject. Time alone can shew which is the better system. I myself have no doubt that it is in general wiser to defer specialization until after a man has taken his first degree, but the drift of recent legislation has been in the other direction.

The curious origin of the term tripos has been repeatedly told, and an account of it may fitly close this chapter. There were three principal occasions on which questionists were admitted to the degree of bachelor. The first of these was the *comitia priora* held on Ash-Wednesday for the best men in the year. The next was the *comitia posteriora* which was held a few weeks later, and at which any student who had distinguished himself in the quadragesimal exercises subsequent to Ash-Wednesday had his seniority reserved to him. Lastly, there was the *comitia minora*, or the general bachelor's commencement, for students who had in no special way distinguished themselves. In the fifteenth century an important part in the ceremony on each of these occasions was taken by a certain "ould bachilour," who as the representative of the university had to sit upon a three-legged stool or tripos "before Mr Proctours" and test the abilities of the would-be graduates by arguing some question with the "eldest son," who was the senior and representative of them. To assist the latter in what was generally an unequal contest, his "father," that is, the officer of his college who was to present him for his degree, was allowed to come to his assistance.

The ceremony was a serious one, and had a certain religious character. It took place in Great St Mary's Church, and marked the admission of the student to a position with new responsibilities, while the season of Lent[1] was chosen with a view to bring this into prominence. The puritan party ob-

[1] Grave scandal was caused at Oxford by a custom of giving suppers after the quadragesimal exercises for the day were over, and this even in

jected to the observance of such ecclesiastical ceremonies, and in the course of the sixteenth century they converted the proceedings into a sort of licensed buffoonery. The part played by the questionist became purely formal. A serious debate still sometimes took place between the father of the senior questionist and a regent master, who represented the university; but the discussion always began with an introductory speech by the bachelor, who came to be called Mr Tripos just as we speak of a judge as the bench or of a rower as an oar. Ultimately the tripos was allowed to say pretty much what he pleased, so long as it was not dull and was scandalous. The speeches he delivered or the verses he recited were generally preserved by the registrary, and were known as the tripos verses: originally they referred to the subjects of the disputations then propounded. The earliest copies now extant are those for 1575.

The university officials, to whom the personal criticisms in which the tripos indulged were by no means pleasing, repeatedly exhorted him to remember "while exercising his privilege of humour, to be modest withal." In 1740, says Mr Mullinger[1], "the authorities after condemning the excessive license of the tripos announced that the comitia at Lent would in future be conducted in the senate-house; and all members of the university, of whatever order or degree, were forbidden to assail or mock the disputants with scurrilous jokes or unseemly witticisms. About the year 1747–8, the moderators initiated the practice of printing the honour lists on the back of the sheets containing the tripos-verses, and after the year 1755 this became the invariable practice. By virtue of this

"the holy season of Lent." Bachelors detected in so acting were liable to immediate expulsion: but as a concession to juvenile weakness the sophister was allowed to give an entertainment in the previous term provided the expenditure did not exceed sixteen-pence. See vol. II. p. 453 of *Munimenta academica*, by Henry Anstey, in the Rolls Series, London, 1868.

[1] Mullinger's *Cambridge*, pp. 175, 176.

purely arbitrary connection these lists themselves became known as the tripos; and eventually the examination itself, of which they represented the results, also became known by the same designation."

A somewhat similar position at the *comitia majora* (or congregation held on Commencement-day) to that of the tripos on Ash-Wednesday was filled by the prævaricator or varier, who was the junior M.A. regent of the previous year, or his proxy. But he never indulged in as much license as the "ould bachilor," and no determined effort to turn that ceremony into a farce was ever made.

The tripos and prævaricator ceased to recite their speeches about 1750, but the issue of the verses by the former has never been discontinued. At present these verses are published on the last day of the Michaelmas term, and consist of four odes, usually in Latin but occasionally in Greek, in which current events or topics of conversation in the university are treated satirically or seriously. They are written for the two proctors and two moderators by undergraduates or commencing bachelors, who are supposed each to receive a pair of white kid gloves in recognition of their labours. Since 1859 the two sets, corresponding to the two days of admission, have been printed together on the first three pages of a sheet of foolscap paper. On the fourth page the order of seniority of the honour men of the year is printed crosswise in columns, the sheet being folded into four parts, so that all the names can be read without opening the page to more than half its extent.

Thus gradually the word tripos changed its meaning "from a thing of wood to a man, from a man to a speech, from a speech to two sets of verses, from verses to a sheet of coarse foolscap paper, from a paper to a list of names, and from a list of names to a system of examination[1]."

[1] Wordsworth, p. 21.

CHAPTER XI.

OUTLINES OF THE HISTORY OF THE UNIVERSITY.[1]

SECTION 1. *The mediaeval university.*
SECTION 2. *The university from* 1525 *to* 1858.

MY object in writing the foregoing pages was to trace the development of the study of mathematics at Cambridge from the foundation of the university to the year 1858. Some knowledge of the history, constitution, and organization of the university is however (in my opinion) essential to any who would understand the manner in which mathematics was introduced into the university curriculum and the way in which it developed. To a sketch of these subjects this chapter is accordingly devoted. I have made it somewhat fuller than is absolutely essential for my purpose, in the hope that I may enable the reader to realize the life of a student in former times.

[1] The materials for this chapter are mainly taken from the *University of Cambridge* by J. Bass Mullinger, Cambridge, (vol. I. to 1535), 1873, (vol. II. to 1625), 1884; the *Annals of Cambridge* by C. H. Cooper, 5 vols., Cambridge, 1842—1852; *Observations on the statutes* by George Peacock, London, 1841; the collection of *Documents relating to the university and colleges of Cambridge*, issued by the Royal Commissioners in 1852; and lastly the *Scholae academicae* by C. Wordsworth, Cambridge, 1877. For the corresponding references to Oxford I am mainly indebted to the *Munimenta academica*, by H. Anstey, Rolls Series, London, 1868, and to a *History of Oxford to* 1530, by H. C. M. Lyte, London, 1886. The works of Peacock, Mullinger and Lyte contain references to all the more important facts.

The history of the university is divisible into three tolerably distinct periods. The first commences with its foundation towards the close of the twelfth century, and terminates with the royal injunctions of 1535. This was followed by some thirty or forty years of confusion, but about the end of the sixteenth century the university assumed that form and character which continued with but few material changes to the middle of this century. Most of its members would, I think agree that a fresh departure in its development then began, the outcome of which cannot yet be predicted.

The mediaeval university.

Cambridge, like all the early mediæval universities, arose from a voluntary association of teachers who were exercising their profession in the same place. Of the exact details of its early history we know nothing; but the general outlines are as follows.

A university of the twelfth or thirteenth century usually began in connection with some monastic or cathedral school in the vicinity of which lecturers had settled. As soon as a few teachers and scholars had thus taken up their permanent residence in the neighbourhood they organized themselves (but in all cases quite distinct from the monastic schools) as a sort of trades union or guild, partly to protect themselves from the extortionate charges of tradesmen and landlords, partly because all men with a common pursuit were then accustomed to form such unions. Such an association was known as a *universitas magistrorum et scholarium.* A universitas scholarium, if successful in attracting students and acquiring permanency, always sought special legal privileges, such as the right of fixing the price of provisions and the power of trying legal actions in which their members were concerned. These privileges generally led to a recognition, explicit or implicit, of the guild by the crown as a *studium generale*, i.e. a body with power to grant degrees which conferred a right of teaching anywhere within the kingdom. The university was frequently incorporated at or about the same time. It was still only a local corporation, but it entered on its third and final stage of development when it obtained recognition, explicit or implicit, from the pope (or emperor). This gave its degrees currency throughout christendom, and it thenceforward became a recognized member of a body of closely connected corporations. Such is the general outline of the history of a mediæval university. In later times the title of university was confined to degree-granting bodies, and any other place of higher education was termed a studium generale.

The records and charters of the university of Cambridge were burnt in 1261, in 1322, and again in 1381. We must therefore refer to the analogy of other universities, and particularly of Paris (which was the typical mediæval university, and was taken as a model by those who first organized Oxford and Cambridge), to obtain an idea of its early history, filling in the dates of the various steps in its development by means of allusions thereto in trustworthy authorities.

It seems almost certain that there was no university at Cambridge in 1112, when the canons of St Giles's moved from the church of that name to their new priory at Barnwell. It is also known that the university existed in its first stage, (i.e. as a self-constituted and self-governing community), in the year 1209, since several students from Oxford migrated in that year to the university of Cambridge. At some time before the latter date, and probably subsequent to 1112, one or more grammar-schools were opened in Cambridge, either under the care of the monks at Barnwell priory, or of the conventual church at Ely, or possibly of both authorities. The connection between these schools and the beginning of the university has always appeared to me to be a singularly interesting historical problem, though it has hitherto attracted but little attention.

Most critics consider that the university of Paris arose from the audiences that came together to hear William of Champeaux lecture on logic in 1109, or his pupil Abelard on theology some thirty years later; and that these lectures were delivered with the sanction of the chapter of Ste. Geneviève. It is generally believed that the university of Oxford arose in a similar way from the students who were attracted there to hear the lectures of Robert Pullen on theology in 1133, and of Vacarius on civil law in 1149; and that as the monks of St Friedeswyde's were probably French, the lectures were given in their house and by their invitation. Paris and Oxford were important towns, and not unnaturally became universities. Cambridge, however, was a small village. In 1086 it only contained 373 hovels grouped round St Peter's church, while about half a mile off were a few cottages clustered round St Benet's Church; and in 1174, after being burnt to the ground, it was only partially rebuilt. It is thus at first sight difficult to see why lecturers should have settled there, and the analogies of other universities throw but little light on it. I suspect the explanation is that students were attracted in the first instance by the great fair held once every year at Stourbridge, which is an open common lying within the boundaries of the borough.

The village of Cambridge was situated at the end of a promontory which projected into the fens, and commanded the northernmost ford by which the eastern counties could communicate with the midlands. Away to the Wash stretched a vast succession of watery fens, across which a stranger could scarcely hope to pass in safety save at the end of a dry summer or after a long frost. The position was thus an important one, both strategically and commercially; and the annual fair at Stourbridge became one of the two great centres of trade for northern and central Europe[1]. Thither the merchants from Germany and the Low countries came by boat from Bishop's Lynn up the Ouse and Cam to exchange their goods for the wool and horses from the western counties and midland shires; and miles of tents

[1] The other great mediæval fairs were Leipzig and Nijnii Novgorod. Stourbridge, though now a mere shadow of its former self and yearly diminishing in importance, is still one of the largest fairs in England.

and booths were put up in streets according to elaborate rules, which at a later time were regulated by act of parliament.

Thus for a month in the year many thousands of travellers were brought to Cambridge, and led, I conjecture, to the establishment of a universitas scholarium, for which the monks and more advanced students of the grammar-schools supplied part of the audiences. It is noticeable that until a few years ago doctors were required to wear scarlet when the fair[1] was proclaimed—thus putting that ceremony on a level for university purposes with the five or six great feasts of the church. Even as late as Newton's time it was apparently an important mart for scientific books and instruments (see pp. 52, 53).

Whatever was the cause of its location at Cambridge the university existed in 1209; and from an allusion[2] in some legal proceedings in 1225 to the chancellor of the university, and from the fact that when in 1229 Henry III. invited French students to leave Paris and settle in England the majority preferred to come to Cambridge, it is clear that it was then an organized and well-known university.

In 1231 Henry III. gave to the university jurisdiction over certain classes of townsmen; in 1251 he extended it so as to give exclusive legal jurisdiction in all matters concerning scholars, and finally confirmed all its rights in 1260. These powers were granted by letters and enactments, and the first charter of which we now know anything was that given by Edward I. in 1291. It was, however, the custom at both universities to solicit a renewal of their privileges at the beginning of each reign (an opportunity of which they often took advantage to get them extended), and it is possible that the dates here given may be those of the renewals of the original charters which, as stated above, were burnt in the fourteenth century.

[1] A collection of references to the fair will be found in pp. 153—165 of the *Life of Ambrose Bonwicke* edited by J. E. B. Mayor, Cambridge, 1870.

[2] Record office, *Coram Rege Rolls*, Hen. III. nos. 20 and 21.

The university was recognized by letters from the pope in 1233, but in 1318 John XXII. gave it all the rights which were or could be enjoyed by any university in Christendom. Under these sweeping terms it obtained, as settled in the Barnwell process 1430, exemption from the jurisdiction both of the bishop of Ely and the archbishop of Canterbury. A survival of this papal recognition, which involved a right of migration, still exists in the customary admission of a graduate of Oxford or Cambridge to an *ad eundem* degree at the other university. The singular privilege of conferring degrees possessed by the archbishop of Canterbury is also derived from the position of the pope as the head of every university in Christendom.

It may be interesting if I add the corresponding dates for Paris and Oxford, since the mediæval histories of the three universities are closely connected. The university of Paris was formed at some time between 1100 and 1169; legal privileges were conferred by the state in 1200; and its degrees were recognized as conferring a right to teach throughout Christendom in 1283. The university of Oxford was formed at some time between 1149 and 1180; legal privileges were conferred by the state in 1214; and its degrees were recognized by the pope in 1296. The university of Cambridge, as I have just explained, was formed at some time between 1112 and 1209; legal privileges were conferred by the state in 1231; and its degrees were recognized by the pope in 1318. Two other mediæval universities rival Paris in antiquity: these were the legal school at Bologna and the medical school at Salerno, but at these the education was technical rather than general.

The characteristic feature of these five mediæval universities—Paris, Bologna, Salerno, Oxford, and Cambridge[1]—is that they thus *grew* into the form they ultimately took. They were recognized by the state and church, but they were not, like the later universities, created by a definite act or charter.

A mediæval university was at first formed of a collection

[1] They are probably the five oldest universities in Europe.

of teachers and pupils with hardly any pretence of organization. So loose was the connection of its members with one another that there was a constant series of secessions. These secessions play a much smaller part in the history of Oxford and Cambridge than in that of the continental universities, as after 1334 the English universities imposed an oath on their graduates never to teach as in a university anywhere in England except at Cambridge and Oxford, "nor to acknowledge as legitimate regents those who had commenced in any other town in England[1]." It must be remembered that the two universities were very closely connected, and that till 1535 a certain proportion of the students divided their time between the two.

It is probable that at the beginning of the thirteenth century there was no code of rules at Cambridge for the guidance of its members. The ancient statutes are undated, but there is every reason to believe that the constitution of the university in the fourteenth century, which is described in the following pages, only differed in details from that which was in practical force during most of the preceding century.

The governing body of the university was termed the regent-house, and it was at first strictly confined to those graduates who were actively engaged in teaching. In the twelfth and thirteenth centuries the final degree of master was merely a license to teach: no one sought it who did not intend to use it for that purpose and to reside[2], and only those who had a natural aptitude for such work were likely to enter

[1] Peacock, Appendix A, xxviii; *Munimenta academica*, 375. At Oxford until 1827 every newly-created master had also to swear that he would never consent to the "reconciliation of Henry Symeon." Henry Symeon is said to have been a master of arts who obtained an office in the reign of King John (1199—1216) by representing that he was only a bachelor. For this offence the implacable university held him up for over 600 years to the obloquy of every successive generation. Peacock, A., xxiii; *Munimenta academica*, 432, 473; Lyte, 214.

[2] A survival of this idea exists in the technical description of a doctor of divinity at Oxford and Cambridge as *sacrae theologiae professor*.

so ill-paid a profession. It was thus obtainable by any student who had gone through the recognized course of study and shewn he was of good moral character. Outsiders were also admitted, but not as a matter of course. By the beginning of the fourteenth century students began to seek for degrees without any intention of teaching; and in 1426 the university of Paris took on itself to refuse a degree to a student—a Slavonian, one Paul Nicolas—who had performed the necessary exercises in a very indifferent manner. He took legal proceedings against the university to compel them to grant the degree, but their right to withhold it was established[1], and other universities then assumed a similar power. He was, I believe, the first student who was "plucked."

The degree gave the right to teach, but after about 1400 the university only granted it on condition that the new master should lecture in the schools of the university for at least one year. Many of those who had ceased to do so were however still resident and engaged in the work of the university; and in course of time heads of hostels, various executive officers, and finally all graduates who had ceased to teach, formed a second assembly called the non-regent house, whose consent was necessary to the more important graces. The two houses taken together formed the senate of the university.

The constitution was thus rendered singularly complex. Some matters were decided by the regents alone, others by the concurrence of both houses voting separately, others by both houses sitting and voting together, and lastly, others by both houses sitting together but with the right of voting confined to the regents[2]. Finally, every measure had to be approved by the chancellor.

The executive of the non-regent house was vested in the two scrutators[3]. But the proctors (sometimes also called rectors) were the two great officers of the university: they

[1] See Bulæus, vol. v. p. 377.

[2] *Statuta antiqua*, 2, 21, 50, 71, 163.

[3] Peacock, 21 *et seq.*

acted as the executive both of the regent-house and of the whole university, and together were competent to perform the duties of the chancellor in case of an emergency. Even the power of veto possessed by him could be challenged if they thought fit; and on their initiative the whole university assembled in Great St Mary's could override the chancellor's veto, or even expel him from his office. It was the proctors as representing the regents (and not the chancellor) who conferred degrees.

The chancellor was chosen biennially by the regents, and acted as head of the university during his tenure of the office. He was always a resident, and it was not until the election for life of Fisher in 1514 that the office became honorary. It is possible that at first the chancellor represented the bishop of Ely, with whose sanction or under whose protection the university had originated, and from whom was derived the power of excommunication[1], which was freely used against troublesome students. The chancellor was however quite independent of the bishop; and so jealous was the university of any possibility of episcopal interference that any official or nominee of the bishop was absolutely ineligible for the office.

The other officers of the university were the taxors, who fixed the rent of hostels and lodgings, and in conjunction with two burgesses determined the price of eatables sold in open market, and four or five beadles who attended on the officers of the university: of the latter two are still retained as the esquire bedells.

It may be added that so soon as a master of arts became a non-regent he was unable to become a regent again except with the consent[2] of the chancellor and the regent-house, a consent which was by no means always given.

Besides these houses the teachers in arts, law, divinity, &c. were constituted into separate faculties, but probably without

[1] Peacock, B., LXV.
[2] *Statuta antiqua*, 11, 144.

legislative powers: the faculty of arts is considerably older than the others[1].

It is probable that at first the university possessed no buildings or appurtenances. Lectures were given in barns, private rooms, or in any place where shelter could be obtained; while congregations of the university and formal meetings were generally held in Great St Mary's Church. At some time before 1346 the university obtained a room or rooms in which exercises could be performed: these were situated in Free-school lane, and were possibly identical with the glomerel schools[2]. The divinity school was commenced in 1347 and opened in 1398; and the art and law schools were added in 1458. The former is now included in the library, and is underneath the present catalogue room (which is itself the old senate-house of the university). The quadrangle was finished in 1475[3]. Most of the colleges and monasteries had libraries

[1] Almost all the above remarks are applicable to Paris and Oxford. The early history of the former has been investigated with great care in *Die Universitäten des Mittelalters bis* 1400, by P. H. Deinfle, Berlin, 1885; and the chief facts connected with it are given in Bulæus. Materials for the history of the university of Oxford exist in great abundance, but I know of no work on it of the same character as that of Deinfle on Paris, or Mullinger on Cambridge.

[2] Mullinger, I. 299, 300. The earliest buildings at Oxford were erected in 1320. (Lyte, 68, 99.)

[3] The following account of the buildings surrounding the eastern quadrangle of the library is taken from the *Cambridge university reporter* of Oct. 20, 1881 (pp. 62, 63). "The northern building, which had the school of theology on the ground-floor, and the 'capella nova universitatis,' or, as it would now be called, the senate-house, on the first floor, was finished about 1400. The west side, which had the school of canon law on the ground-floor, and the 'libraria nova' on the first floor, had been commenced in 1440, but was not completed until 1458. The south side, which had the schools of philosophy and civil law on the ground-floor, and some other schools, together with a library, on the first floor, was erected between 1458 and 1467. The narrow building that joined the north and south sides together, and formed a west front, continuous with the eastern gables of the north and south sides, was erected between 1470 and 1475. The ancient aspect of this quadrangle is shewn in

of their own, but the university or common library was not built till 1424.

The question of how suitable provision should be made for the board and lodging of the students was however far more pressing than that of providing accommodation for the corporate life of the university. The town was originally only a collection of unhealthy cottages, and unlike Paris and Oxford had no importance except that derived from the presence of the university. The character of the accommodation offered did not however prevent the townsmen from utilizing their monopoly to make extortionate charges; and almost the first act of the university of which we know anything was to attempt to find a remedy for the evils and dangers to which the lads who flocked to the university were thus exposed. In 1231 a rule was made that every scholar must place himself under the tuition of some master[1]: and in 1276 the university, in virtue of powers conferred by the crown, passed a grace that no lodging-house keeper or teacher was to receive a scholar unless the latter "had a fixed master within fifteen days after his entry into the university[2]." No record of this tutorial relation was kept by the university, but at stated periods the masters attended in the schools and read out the roll of their

Loggan's print, taken about 1688. The porch and staircase at the N.W. angle, together with the west wall as far as the northernmost buttress on that side, was taken down in 1714, in order to make a room on the first floor large enough to receive Bishop Moore's library. At the same time the windows, of which there was originally only one, of three lights, between each pair of buttresses, were replaced by the existing round-headed ones. Shortly after, in 1727, the present senate-house being completed, the old 'capella universitatis' was absorbed into the library. The classical building, which now replaces the central block on the east side, was begun in 1754, the style being selected in order to make it harmonize with the senate-house. The old divinity school on the groundfloor of the north side was taken into the library in 1856. These various changes have utterly destroyed the ancient character of the quadrangle."

[1] Cooper, I. 42.

[2] *Statuta antiqua*, 42.

own pupils[1]. There was no formal matriculation of students until the year 1543[2].

The university also took steps to encourage the resident masters to open hostels or boarding-houses, and until the sixteenth century the majority of the students lived in these houses. One of the earliest of the extant statutes[3] of the university gives the detailed rules which the university laid down about the year 1300 for regulating the hiring of these hostels. It illustrates how completely the university was then the dominant power in the town, that if a master of arts wished to take any particular house for a hostel and could give security for the rent the university turned the owner out[4].

Another way of meeting the difficulty was by the establishment of colleges, the idea of which was borrowed from Paris and Oxford. The earliest to be established was that which is now known as Peterhouse in or before 1280. At first this and other similar foundations were designed to house and support a master with certain fellows and scholars (to give them their modern designations) only, but not pensioners or ordinary students.

Another danger of a different kind existed in the constant efforts at proselytizing by the religious orders. In the course of the thirteenth century all the great monastic orders established houses in Cambridge where food, shelter, the use of a library, and assistance were offered to all who would join the order. The number of these houses shew that the reputation of the university must have been considerable. The Augustinian canons were already established at Barnwell, but they enlarged their abbey till it became one of the wealthiest in the kingdom. The Franciscans built a house in 1224, and shortly

[1] *Cambridge documents*, I. 332. Lyte, 198.

[2] Mullinger, II. 63.

[3] It is printed at length in Mullinger, I. 639, and a translation is given on pp. 218—220.

[4] See vol. I. p. 65 of Cooper's *Annals* on a case which happened in 1292: it is evident from the references that the university was legally entitled to exercise the power.

afterwards erected the magnificent church and monastery which formerly stood on the site of Sidney. By the middle of the thirteenth century representatives of nearly all the orders were living in Cambridge. Among others the Carmelites had occupied the site of Queens'; the Dominicans the site of Emmanuel; the Augustinian friars the site of the new museums; the Benedictines the site of Magdalene; the White canons the site of Peterhouse Lodge; and the brethren of St John the site of the college of that name[1].

Now the university, though it was closely connected with the regular clergy and though the majority of its members were even in orders, was still essentially a secular institution. It was natural, therefore, that this crowd of monks, who merely became masters of the university in order to recruit among its junior members, should be regarded with great suspicion. The successful ruse by which in 1228 the Dominicans had temporarily obtained the entire control of the university of Paris gave warning of what was designed, but with that toleration which has always been a marked feature in Cambridge life an open rupture was avoided—the monks were admitted to degrees so long as they conformed to the regulations of the university, and by courtesy one was always elected on the caput[2] (see p. 245).

The university, however, never ceased to be on its guard against these "foreigners who," so ran the phrase, "cajoled lads before they could well distinguish betwixt a cap and a cowl." In 1303 two of them, Nicholas de Dale and Adam de Haddon, insisted that the rights of their respective monasteries were paramount to all privileges of the university[3]. They were accordingly expelled; but in 1306[4] the university allowed monks to proceed to degrees in divinity without having previously incepted in arts. Instead of accepting this decision

[1] Mullinger, I. 138, 139, 564.
[2] *Statuta antiqua*, 4; Peacock, 21.
[3] Peacock, 26.
[4] Peacock, 33.

as a favor and concession the monks treated it as a sign of their triumph, and in 1336 a grace had to be passed forbidding the friars to receive into their orders any scholar under the age of eighteen. Oxford passed a similar statute in 1358. Under pressure from Rome these statutes were subsequently repealed, but in 1359 the university passed a grace by which only two friars from each house were allowed to incept in the same year[1], which sufficiently served to protect the university from excessive proselytizing.

The establishment of these numerous and powerful bodies had however another and more lasting effect. Although the monks and friars were nominally members of the university, they were divided from the rest of the masters on nearly every question of policy, and thus acted as a counterpoise to the overwhelming power of the university in local matters. They were also wealthy, and materially increased the prosperity of the town, so that by 1300 the mayor and burgesses formed a well-organized corporate body. In that year the total population of the university and town was about 4000[2], but except at the time of the annual Stourbridge fair there does not seem to have been any considerable trade, save that arising from the supply of the needs of the university and the monasteries.

The statements about the number of students at the mediæval universities must be received with considerable caution. They represent vague impressions rather than the result of an accurate census. It must also be recollected that it was customary to reckon as members of the university all servants and tradesmen whose chief employment was in connection with students, while the fact that the average student spent at least seven years at the university before he became a master, and generally twenty years or more if he aspired to become a doctor (after which he probably still resided for some years), caused the university to be largely composed of permanent residents of every age from 12 to 40.

[1] *Statuta antiqua*, 163, 164. Peacock, xliii; Mullinger, I. 263.

[2] Cooper, I. 58.

The question has been very carefully considered by M. Thurot[1], who comes to the conclusion that the total number of students at Paris never rose much above 1500 nor of regents above 200. I think I should probably not be far wrong if I estimated the total number of masters and students (exclusive of monks) at Cambridge during the thirteenth, fourteenth, and fifteenth centuries as varying between 500 and 1000. The numbers at Oxford in the thirteenth century were perhaps about 700; in the fourteenth century probably nearly 2000; in the fifteenth century the university is described as "wholly deserted," perhaps the total number then did not exceed 200 or 300. I ought to add that all these numbers are considerably less than those usually given, but the latter probably include servants and tradespeople. Peacock says[2] that the number of regent-masters created at Cambridge in each year [I presume in the fifteenth century] averaged about 40; and that of bachelors in law about 15. This, as far as I can judge, will give a result not very different from that which I had independently arrived at.

The question as to the social position of the students in mediæval times is a difficult one[3]. The balance of opinion is that a large majority were poor, and it is certain from several of the ancient statutes that poverty was not uncommon[4]. On the other hand, a considerable minority must have been wealthy. The grace, to which allusion was made in chapter VIII., by which any incepting master was forbidden to spend in presents and dinners, on the occasion of taking his degree, what would now be equivalent to £500, would have been absurd if there were no wealthy men at the university. Moreover it is clear from internal evidence, that Richard II. in framing the statutes of King's Hall (which had been founded by Edward II.

[1] See pp. 32, 42 of *De l'organisation de l'enseignement au moyen âge*, by C. Thurot, Paris, 1850. See also *Munimenta academica*, p. xlviii.

[2] *Observations*, 33.

[3] Mullinger, I. 345, note.

[4] See Cooper, I. 245, 343.

and Edward III., and is now a part of Trinity College), expressly designed it for wealthy and aristocratic students[1]. All regulations about poverty were erased from its rules, while in place of them various sumptuary and disciplinary regulations were inserted. Among these I notice that the daily expenditure of food for each student was not to exceed 1*s.* 2*d.* a week, which would be worth now say about 14*s.* or 15*s.* and, was nearly half as much again as at Gonville Hall. Other rules were that students should not keep dogs in college, or play the flute to the annoyance of their neighbours. The additional provision that no one should practise with the cross-bow in the courts or walks of the college must commend itself to every one of mature age. A tradition that the society laid down a rule that no student should strike a fellow, or under any circumstances the master, is suggestive that its members were not wholly devoted to study. In the fifteenth century no one was admitted who was not *bene natus.*

I think therefore we may safely say that the students were drawn from all classes and ranks in the kingdom, but that a large proportion were poor.

I may perhaps be pardoned for adding a few words on the social side of the life of a mediæval student. The majority of the students and all the wealthier ones resided in hostels[2]. Some of these houses no doubt contained all the comforts which were then customary, but no account of life at a hostel is now extant. It would seem, however, that there was usually a common sitting-room or hall; and at the better hostels a lad could hire a bedroom for his sole use, the rent of which varied from 7*s.* 6*d.* to 13*s.* 4*d.* a year[3]. The total expenditure of the son of a well-to-do tradesman at Oxford in the reign of Edward III. came to £9. 10*s.* 8*d.*; board was charged at the rate of 2*s.*

[1] Mullinger, I. 252—254.

[2] See Lever's sermon at St Paul's Cross, preached in 1550: Arber's edition, p. 121.

[3] *Munimenta academica,* 556, 655.

a week, tuition at 26*s.* 8*d.* a year, and clothes cost 20*s*[1]. In 1289 the allowance to two brothers de la Fyte was half-a-mark each per week, which was raised in their second year of residence to 35 marks a year: besides this bills for certain necessary expenses, which seem to have averaged nearly £5 a year for each of them, were paid by the king. This scale of allowance was exceptionally high, as the boys were well connected, and protected by the king: they had a manservant to themselves. At the other end of the social scale two poor lads named Kingswood were sent by bishop Swinfield to Oxford in 1288, and the bills for both of them for forty weeks' residence came to £13. 19*s.* 2*d.*[2] From these and similar facts it would seem that a student could hardly support himself on less than £9 a year, and that anything beyond £15 a year was a handsome allowance. If these totals be multiplied by 12 or 13 they will represent about their equivalents in modern value.

The colleges, except King's Hall, were intended for poor students, but compared with those of Paris seem to have been fairly comfortable, and indeed for that age luxurious. Every student swore obedience to the college authorities, and it was rigidly enforced with birch and rod. The younger students slept three or four in a room, which also served as study, but was more often than not unwarmed. There was a dining hall, in which on great occasions a fire was lit. Here meals were served, namely, dinner about 10 a.m. and supper about 5 p.m.; meat being apparently provided on each occasion, except in Lent. The colleges generally required their members to speak nothing but Latin (or in a few cases French) in hall and on all formal occasions except the great festivals of the church. In the evening mock contests were held in the hall, by which students were practised for the acts they had to keep in the schools. There was usually an attic fitted up as a library where students could find the text-books of the day, and

[1] The accounts of the guardian of Hugh atte Boure, quoted in Riley's *London*, p. 379.

[2] The authorities are quoted in Lyte, 93.

from which a fellow could borrow books: this use of a library was one of the most highly valued privileges of college life[1].

The disciplinary rules of the colleges were naturally stricter than those in force in the hostels. Until a student of a college became a bachelor he was not allowed to go out of college bounds unless accompanied by a master of arts. A bachelor had much the same freedom as an undergraduate now-a-days, except that he generally had but one room, which he had to share with another man, and only a fellow of considerable standing had a room to himself. Allowances were conditional on residence, but were generally sufficient to supply all the necessaries of a student's life. The master was absolute within the college: a fatal defect in organization, for a single incompetent master could destroy the progress of centuries, as every mediæval college in succession found to its cost[2].

The amusements[3] of the students were much what we should expect from English lads. Contests with the cross-bow were common, and cock-fighting—at any rate in the hostels—was a usual amusement. To the more adventurous student the opportunity of a fight with the townsmen was always open. As far as we can judge at this distance of time the university authorities in their dealings with the town were arrogant and exasperating, but always kept within the law; and technically in all the serious riots the townsmen were in the wrong. The riots of 1261, 1322, and 1381 were particularly violent, and the townsmen not only committed outrages of every kind, but burnt some of the hostels, and all the charters and documents of the university as well as of such colleges as they were able to sack. After the last of these riots the government confiscated the liberties of the town, and bestowed them on the chancellor, in whom they remained vested till the reign of Henry VIII. To this stringent measure the subse-

[1] Mullinger, I. 366—372.

[2] See for example Mullinger, I. 424.

[3] Mullinger, I. 373, 374.

quent prosperity of the university (and so indirectly of the town) was largely due. The ill feeling which existed at every mediæval university between town and gown was intensified at Cambridge by the fact that the fishing in the river was unusually good, and belonged absolutely to the mayor and corporation, who refused to allow university men to fish in it under any circumstances. Such a right could not be enforced without considerable friction, and as the university claimed and exercised exclusive jurisdiction to try cases where their own members were concerned, the dispute was complicated by differences of opinion on the evidence requisite to prove a trespass or assault[1].

Besides these amusements there was rarely a year in which some tournament or form of sport was not held in the immediate neighbourhood, and like the fair at Stourbridge gave opportunity for plenty of adventures, as well as the interesting spectacle of bear and bull baiting. The prohibitions in the statutes of New College, Oxford, of dice and chess as instruments of gaming imply that they were constantly used. Among the more wealthy members of the university tennis, cock-fighting, and riding seem to have been especially popular; but many of the college statutes enjoin that a daily walk with a companion, and conversation "on scholarship or some proper and pleasant topic" should if possible be enforced.

Lastly, it should be added that local ties and prejudices were very strongly maintained. Students born anywhere south of the Trent formed one "nation," while those born to the north of it formed another. These nations took opposite sides on every question; thus when Occam, who was a southerner, advocated nominalism, the northerners at once adopted the

[1] Finally, in despair of obtaining their rights otherwise, the corporation farmed their powers piscatorial to certain poor men, who it was thought "needing all the money they could obtain would not fail in well guarding that which they had purchased." This ingenious scheme failed, for the poor men shortly petitioned the corporation to cancel the agreement, since "many times had they been driven out of their boats with stones and other like things, to the danger of their bodies."

realistic views of Scotus. They were organized[1] almost like regiments, and the smouldering hostility between them was always ready to break into open riot, which not unfrequently ended in loss of life. So high did local feeling run that most of the college statutes expressly guarded against the favoritism that arose from it by a provision that not more than two or three scholars or fellows born in the same county could be on the foundation at the same time.

The students dressed much like other Englishmen of the same period. Efforts to enforce the tonsure and ecclesiastical robe were not unfrequently made, but seem to have been always evaded. Perhaps knee-breeches, a coat (the cut of which varied at different times) bound round the waist with a belt, stockings, and shoes (not boots) fairly represent the visible part of the dress of an average student at an average time. The dress of a blue-coat boy may be compared with this. To this most students seem to have added a cloak edged or lined with fur, which often found its way into the university chest as a pledge for loans advanced. Girdles, shoes, rings, &c. varied with the fashion of the day.

The earliest inventory of the possessions of a Cambridge student that I can quote is one of the belongings of Leonard Metcalfe, a scholar of St John's College, who was executed in 1541 for the murder of a townsman. All his goods were confiscated to the crown, and therefore scheduled by the vice-chancellor[2]. His wardrobe consisted of a gown faced with satin, an old jacket of tawny chamblet (i.e. silk and hair woven crosswise), an old doublet of tawny silk, a jacket of black serge, a doublet of canvass, one pair of hose, an old sheet or shirt, a cloak, and an old hat. I suppose these were in addition to the clothes he wore when being executed, as the latter were the

[1] See *Statuta antiqua*, 44.

[2] See vol. I. pp. 109, 110 of the *Privileges of the university of Cambridge*, by George Dyer, London, 1824. For corresponding inventories of Oxonians, see *Munimenta academica*, numerous references between pp. 500—663.

perquisite of the hangman. He had besides a coverlet, two blankets (one being very old), and a pair of sheets—but most of these are stated to have been pawned before he went to prison. His furniture consisted of a wardrobe-chest with a hanging lock and key, a piece of carpet, a chair, a knife, and a lute. The table and bedstead were fixtures, and belonged to the college. His books with their respective values were as follows. A Latin dictionary, 1*s.* 8*d.*; Vocabularius juris et Gesta Romanorum, 4*d.*; Introductiones Fabri, 3*d.*; Horatius sine commenti, 4*d.*; Tartaretus super Summulas, 2*d.*; The shepheard's kalender, 2*d.*; Moria Erasmi, 6*d.*; and Compendium quatuor librorum institutionum, 3*d.*; the total value being three shillings and eight-pence, equivalent to rather more than two pounds now-a-days. He had not taken his bachelor's degree, and it is therefore not surprising that he possessed no mathematical works. His total assets were valued at £4. 1*s.* 8*d.*, equivalent to £50 or £60 at the present time. The above list seems fairly to represent the belongings of a mediæval student, except that Metcalfe's library was unusually large.

A gown or some similar distinctive dress has always been worn at Cambridge[1]; but the cut and material varied at different times. Masters wore a square cap, and doctors a biretta, but it is not clear whether any cap was worn by undergraduates. From the original statutes of New College, Oxford, and Winchester School, it seems probable that at that time the students went bareheaded, as they still do at Christ's Hospital. The earliest reference to caps being worn by students as a part of their academical dress occurs in the sixteenth century. The cap then worn was circular in shape and flabby, lined with black silk, with a brim of black velvet for pensioners or black silk for sizars. The square cap for undergraduates was not generally introduced till 1769: the puritan party having objected to it in the sixteenth and seventeenth centuries as a symbol of popery.

The cut of the B.A. hood has not varied from the thir-

[1] See Cooper's *Annals*, vol. I. pp. 156, 157, 182, 215, 355.

teenth century, except that the two ends were formerly sewn together instead of being connected by a string as they are now. In the middle ages it was lined with wool and not rabbit-skin. The shape is different to that of all other universities, as it includes what is called a tippet. The M.A. hood for regents was the same as at present. The hoods of non-regents were of the same shape, but lined with black. The proctors invariably wore the hood squared, as they do now: and the scrutators and taxors had the same privilege[1].

It must be remembered that the mediæval university and colleges were very poor[2]. The members of the latter often found themselves unable to obtain money, even for their daily food, except by selling books or pledging their house. The former had a few scholarships, the earliest of which was founded in 1255, and possessed a few funds for the purpose of loans. Every separate bequest or gift was for simplicity of accounts kept in a separate chest, and some of these coffers are still preserved in the registry. The name has also been retained as a synonym for the university treasury.

The development of the university throughout the middle ages seems to have been one of steady, uniform progress. This was partly due to its own merits, but partly to the gradual deterioration of the monastic schools. There was no sudden outburst of prosperity, such as that which in the fourteenth century made Oxford the most celebrated seat of learning in Europe, but neither was there any collapse such as that which in the fifteenth century left Oxford almost deserted; though the numbers at Cambridge do not seem to have increased during that century.

[1] The above account is summarized from pp. 454—543 of *University life in the eighteenth century*, by C. Wordsworth, Cambridge, 1874.

[2] Even now the corporate revenue of the university proper (as distinguished from the colleges) is less than £2,500 a year. I suppose very few people realize how pressed for means is the university, and that it is only by contributions from the colleges (out of property which was really left for other purposes) that the university contrives to balance its accounts. The much greater wealth of the sister university has largely contributed to the idea that the university of Cambridge is also wealthy.

The university from 1525 *to* 1858.

The close of the fifteenth century was marked by the commencement of schools of science and divinity. A similar development was general throughout Europe, but it was some years before the English universities felt the full force of the movement. The intellectual life at Oxford during the middle ages had been far more vigorous and active than that at Cambridge, and in literature (though probably not in science and divinity) the renaissance in England had commenced about the year 1440 at Oxford. The logicians there bitterly opposed the new movement, and succeeded in temporarily stopping it. The consequence was that the revival of the study of literature in England was mainly effected at Cambridge. The effects of this preeminence in the sixteenth century lasted long after the immediate causes had ceased to act, and until the close of the eighteenth century the literary and scientific schools of Cambridge were superior to those of Oxford.

It was to Fisher, and subsequently to Erasmus, that Cambridge owed the creation of its literary schools, which originated about the year 1510. I think, however, that during the preceding century—in fact since the suppression of the Lollard movement by Archbishop Arundel on his visit in 1401—the drift of opinion in Cambridge had steadily set towards moderate puritanism and the study of science. I suspect that the divergence in the opinions prevalent at Oxford and Cambridge which here first shews itself was due to the fact that the residents at Cambridge were every year brought into contact at the Stourbridge fair with merchants and scholars from Germany, and apparently through them with the Italian universities (especially Padua), while Oxford was a much more self-contained society. It is noteworthy that almost all the Cambridge reformers came from Norfolk, which was in close commercial connection with the Netherlands, and that the literary party in the university were nicknamed Germans.

On the other hand it should be noted that some of the most influential leaders of the renaissance (such as Tonstal, Tyndale, Recorde, and Erasmus) came from Oxford, bringing with them the best traditions of that university; and the rapidly rising reputation of Cambridge was greatly stimulated by those new-comers. So completely successful were the philosophers at Oxford in destroying the study of literature there, that Wolsey was obliged to come to Cambridge, much though he disliked it, to get scholars acquainted with the subject to put on the foundation of his new Cardinal College. The same reason probably explains why some fifty years later the society of Trinity College, Dublin, was at first almost wholly recruited from the members of Trinity College, Cambridge.

The triumph of the Oxford logicians was synonymous with the ascendancy there of the narrow orthodox theological party. Hence the reformation was mainly the work of Cambridge divines. The preliminary meetings in which the general lines of the movement were laid down were all held at Cambridge at the White Horse Inn, where the house of the tutor of King's now stands. The most prominent of these proto-reformers were Barnes, Bilney, Coverdale, Tyndale, and Parker. The prevalent feeling of the university is shewn by the fact that when in 1525 Wolsey ordered the arrest of Barnes the students broke into the room in which the court before which he had been summoned was sitting, and Wolsey had to adjourn the trial to London before he could secure a hostile verdict. Many of the most eminent members of the university, such as Cranmer, Ridley, Latimer, Ascham, and Cheke, did not conceal their sympathy with the reformers. The fall of Wolsey and the rise of Cranmer (who had suggested Henry's divorce) threw the control of the movement entirely into the hands of graduates of Cambridge, and perhaps no more striking evidence of that can be given than the fact that out of the thirteen compilers of the new prayer-book issued in 1549 twelve came from Cambridge, while the litany was prepared

16—2

by Cranmer from the work of Wied and Bucer[1]. On the other hand, all the leaders of the Roman party (save Fisher, who belonged to an older generation) were Oxonians.

The development of the study of classical and biblical literature and of science, and the rise of a critical spirit evoked by the renaissance mark the approaching end of the reign of the schoolmen, and the mediæval curriculum was definitely terminated by the royal injunctions of 1535. In these the king ordered that henceforth no lectures should be given on the sentences or on canon law; but that Greek, Latin, and divinity should be taught in addition to the trivium and quadrivium, and that the scriptures should be read. The university system of teaching by means of the lectures of the regents was essentially bad. To remedy this it was ordered that permanent lecturers should be appointed. At the same time the large number of clergy and others who were living at Cambridge to enjoy the social advantages of the place, without any intention of studying, were ordered to quit it at once if over forty years old[2].

This break-up of the mediæval system of education was followed by a serious fall in the number of students, until in 1545 the entries barely exceeded 30, while at Oxford they sank to 20. So serious did the situation become that the university directed all "useless books" in the university library to be sold; and abolished some of the annual offices in the university, directing that their duties should be performed by the proctors as best they might. In 1535 and 1537 the university even suspended the Barnaby lecturer on mathematics, so that they might appropriate his salary of £4 a year for the benefit of the lecturers on Hebrew and Greek.

After the dissolution of the monasteries, Henry VIII.

[1] Bucer was regius professor of theology at Cambridge, and worked in collaboration with Wied.

[2] Mullinger, i. 630.

personally investigated the position of the universities, and decided that they were doing admirable work in an economical and efficient manner[1]. To promote study he endowed at Cambridge in 1540 five regius professorships (see p. 154).

It was at this time that the colleges began to admit pensioners as well as scholars (see p. 154). The effect on the members of the university was immediate and striking. In 1564 the number of residents had risen to 1267, and in 1569 it was 1630. The corresponding numbers at Oxford were rather less than two-thirds those of Cambridge.

The Edwardian statutes of 1549 were an honest attempt to reorganize the university in a manner suited to the changed conditions of education (see p. 153), but no serious alterations were made in the constitution.

The Elizabethan code of 1570 made numerous changes[2]. That code was mainly designed to effect three things: first, on the advice of Cecil, to make the university directly amenable to the influence of the crown; secondly, on the advice of the bishops, to make it a distinctly ecclesiastical organization, with a view to provide a supply of educated clergy for the realm; and thirdly, probably by command of the queen, to ensure that the best general education for laymen as well as clergy should be obtainable; finally, the better to secure these objects it was decided to offer no direct encouragement to any other work. The university strenuously opposed this limitation of its powers and studies, but without success.

The subjection of the university to the power of the crown was effected by an ingenious artifice suggested, it is believed, by Cecil. From time immemorial the first grace at a congregation was to appoint a committee of five, termed the *caput*, to assist the chairman at that meeting. To prevent objectionable or surprise motions a grace could not be submitted if any member of the caput objected to it. By the new statutes the caput was constituted as a permanent committee, to be elected by the

[1] Mullinger, i. 461.

[2] Mullinger, ii. 222—34.

heads of colleges, doctors, and the two scrutators, and to hold office for a year. Without going into further details it may be said that this gave an absolute veto, and also the whole power of initiating legislation, to an irresponsible committee appointed by the heads: and even then, the vice-chancellor could frustrate all legislation by refusing to summon the committee, as happened in 1751—52. The heads were also directed to nominate two names for the vice-chancellorship, one of whom must be chosen; and consequently since 1586 no one but a head has been elected to that office. Finally, the heads were to act as a council to advise the chancellor on all matters affecting the conduct of students, and were to fix the times and subjects of all exercises and lectures. Besides this each head was given a power of veto on any public act or election in his own college. The rights of the regent and non-regent houses were not directly touched, but practically the heads were made supreme; and as there were but fourteen of them, nearly all of whom were hoping for preferment at the hands of the crown, there was little difficulty in getting their sanction to anything the government wished. The proctors, who were entitled if they wished to set aside both chancellor and caput and to appeal directly to the university, were deprived of most of their powers, and expressly declared to be like all other officers subordinate to the chancellor. Henceforth they were nominated by the colleges according to a certain cycle, and the nomination was conditional on the approval of the heads.

That the old democratic construction was open to grave abuses is evident from the unscrupulous tactics of the puritans at some of the congregations in the spring of 1570. That party were not then strong enough to control the policy of the university, but they were able to block all business and legislation. Several congregations broke up in great disorder, and it was necessary to make the executive efficient, whichever party controlled it. The new oligarchic constitution erred on the other side and almost stifled the independent criticism of the senate. At the same time I should observe that any member of the senate could propose a grace, and, except in times of great excitement, it was usual to allow it to be put to the vote. It will be noticed that by the statutes of 1858 many of the powers of the caput were transferred to a council elected by the resident graduates, which is so far perhaps a reasonable compromise, but against this must be set the fact that the members of the senate have practically been deprived of the power of initiating a grace.

To secure the ecclesiastical character of the university a decree of 1553 was confirmed, by which the subscription of the forty-two articles was required from all those proceeding to the degree of M.A., B.D., and D.D.; and in 1616 this was extended to all degrees.

The commissioners who drafted the Elizabethan statutes of 1570 not only reorganized the constitution of the university but recast the curriculum. Mathematics was excluded from the trivium, and undergraduates were directed to read rhetoric and logic, but the course for the master's degree was left almost unaltered (see p. 156). The necessary exercises for degrees and intervals between them were left as before, except that they were defined rigorously by statute, and no resident could be excused from any of them. The regency of masters was extended to five years, after which a master became necessarily a non-regent. Generally the discipline of the university was made more precise and rigid.

The new statutes recognized the change which had taken place in the system of education by assigning to a regent the duty of presiding over or taking part in the public disputations, and not as formerly that of teaching and reading in the schools. Finally, new statutes could only be made if they in no way interfered with these.

The commissioners saw that the mediæval university had failed to provide teaching suitable for most of its members, and had made no proper provisions for the safety and discipline of the students; and they realized that for the future the efficiency of the university must largely depend on that of the colleges.

They accordingly spent two months in visiting the separate colleges[1]. The chief object of the changes introduced was to secure good discipline and teaching, and decency in public worship[2]. The commissioners entered into such detail as to settle the dress of members of the university for all time to come, and even the private prayers they should use when they got out of bed in the morning.

Some of the provisions of these statutes, such as the regency of five years, the power of veto in all college matters by its master, and possibly the residence of bachelors, were never enforced, and others were constantly broken; but taken as a whole they were accepted by the university and acted on.

Shortly after the Elizabethan statutes came into effect the incomes of the colleges began to rise, partly through their good management of their estates, partly by gifts of their members. It became not uncommon to have a surplus after meeting the expenses of the house, and as the surplus, if any, was divisible among the fellows, a fellowship began to be regarded as a money prize which might serve as a provision for life—an idea which no doubt materially retarded the intellectual life of the university.

The following table, which is as complete as the material at my command permits, will enable the reader to judge of the progress of the university. It gives for the various periods mentioned the average yearly number of matriculations, and the average yearly number of bachelor degrees (exclusive of those of

[1] See the contemporary account published in Lamb's *Documents*, London, 1838 (pp. 109—120).

[2] I think few people realize how intolerant were the extreme puritan party at this time, and how anxious they were to display their principles in such a way as to hurt what they regarded as the prejudices of their contemporaries. As an illustration of the length to which they were prepared to go, I may mention that at Emmanuel (their head-quarters in the university) they took the communion "sittinge upon forms about......& did pull the loafe one from the other......and soe the cupp, one drinking as it were to another like good fellows." (Baker VI. 85—86, quoted by Mullinger.) Had they been more tolerant and courteous I believe they would have triumphed; but their excessive zeal provoked a continual reaction against them and their doctrines.

medicine and theology) which were conferred. The number of undergraduates resident in any year after 1600 may be taken roughly as being four times the number of those who took the B.A. degree in that year. I have added the corresponding numbers for Oxford wherever I could obtain sufficient data, but I have no doubt that the statements about the numbers of matriculations there in the sixteenth and seventeenth centuries (although founded on official data) are incorrect[1].

Period	Cambridge *matriculations*	Oxford *matriculations*	Cambridge *bachelors*	Oxford *bachelors*
From 1501 to 1516	....	....	... 48...	....
,, 1518 ,, 1570	....	....	... 50...	... 43 ...
,, 1571 ,, 1599	....	... 258 (?)	...178...	... 110 ...
,, 1600 ,, 1633	....	... 312 (?)	...229...	... 191 ...
,, 1634 ,, 1666	....	....	...193...	....
,, 1667 ,, 1699	....	... 326 (?)	...185...	... 174 ...
,, 1700 ,, 1733	....	... 297 ...	...151...	....
,, 1734 ,, 1766	....	... 214 ...	...106...	....
,, 1767 ,, 1799	......153 ...	... 241 ...	...114...	....
,, 1800 ,, 1833	......342 ...	... 332 ...	...230...	....
,, 1834 ,, 1866	......447 ...	... 423 ...	...346...	....
,, 1867 ,, 1886	......743 ...	... 693 ...	...565...	....
In 1887	... 1012 ...	... 766 ...	...786...	... 612 ...

There is but little difficulty in describing the life, studies, and amusements of the students of this period. From the

[1] The numbers given for different years are extraordinarily various and bear no relation to the number of B.A. degrees conferred four years later. Thus the matriculations for 1573 and 1575 are returned as 35 and 467 respectively, while the number of B.A. degrees taken sixteen terms (four years) later are given as 97 and 115: the latter are probably correct. In some years the entry is stated as having been larger than is the case now (e.g. the return for 1581 is 829), and it is certain that there was then no accommodation in the colleges for such numbers. We have also good reason for saying that from 1570 to 1620 the number of residents at Oxford was about two-thirds of the corresponding numbers at Cambridge, and thus must have been much smaller than the alleged number of matriculations. I have therefore no doubt that the data are untrustworthy.

close of the sixteenth century there is a constant succession of diaries, and a great mass of correspondence by resident members of the university. The social life of the seventeenth century is described at length by Mullinger (vol. II. chap. v.), and that of the eighteenth century by Wordsworth. It was rougher and coarser than that to which we are accustomed, but it was more civilized and courteous than that of the middle ages.

The most popular amusements of the undergraduates of the upper classes in the seventeenth century seem to have been tennis, cock-fighting, fishing, hawking, hunting, fencing, and quoits (at one time or another). Football also was apparently occasionally played[1]. Students of the lower classes seem to have indulged in a good deal of rough horse-play. The long winter evenings were relieved by plays performed in hall after supper on Saturday and Sunday evenings; and at Christmas every one, young and old, played cards. But with compulsory morning chapel at 6 a.m., and deans who would take no excuse for absence, the hour for bed was earlier than at present.

The usual amusements of the undergraduates of the eighteenth century were tennis, racquets, and bowls: fives and billiards were also occasionally played. There were no athletic clubs[2], and the only organized societies (other than dining clubs) that I know of were those for ringing peals on church-bells and giving concerts. The annual fair at Stourbridge was the meeting-place of nearly every conjurer, mountebank, and company of strolling actors in the kingdom, and for a fortnight provided a perfect surfeit of amusements.

Discipline was stern. The birch rod, which during the seventeenth century and the early half of the eighteenth century hung up at the butteries, was in regular use; and once a

[1] D'Ewes mentions a match in 1620 between Trinity and St John's.

[2] Boat-racing on the river was apparently introduced about 1820, and cricket some twenty or thirty years earlier: it is said that the first public match of cricket in its present form ever played was that of Kent against England in 1746.

week the college dean attended in hall—usually on Thursday evenings—to see that the butler applied it to such youths under the age of eighteen years as had infringed any college rules, or sometimes to any lad who was beginning to shew himself "too forward, pragmatic, and conceited".

At sunset the college gates were locked. All the students however lived in college, and the more popular colleges were so overcrowded that usually three or four men had to share a room. Except at Trinity, where most of the students were sons of county squires or parsons, the bulk of the students came from what is called the lower middle class, but there was a fair sprinkling of members of the aristocracy who lived apart from the rest of the community. The expense to the son of a county squire seems to have been equivalent to from £180 to £220 a year; to a fellow-commoner about £330 a year. The servants of the college, porters, cooks, &c. were mostly sizars, who received education, board, and lodging in return for their services.

The hour of dining gradually grew later[1]. In 1570 it was at 9·0, or at Trinity at 10·0. By 1755 it had got shifted to noon. In 1800 it was at 2·15 at Trinity, and at 1·30 at most of the other colleges; and the senior members of the university began to complain that the afternoon attendance at the schools was in consequence much diminished. A few years later dinner was usually served at 3·0, but until 1850 the hour did not, I think, get later than 5.0. Since then the same movement has gone on, and now (1889) dinner at Trinity is at 7.30.

The main outlines of the history of the university under the Elizabethan code are probably well known to most of my readers. The leading features are connected with the history of the theological school, the rise of the mathematical and Newtonian schools, and finally the outburst of activity in all departments of knowledge which preceded the grant of the first Victorian statutes.

The supremacy of the Cambridge school of theologians

[1] Wordsworth, 119—129.

remained unbroken till the death of James I.; and it may be illustrated by the fact that no less than four out of the five delegates from Britain to the synod of Dort in 1618 came from Cambridge. Its influence in the country was then destroyed by the rise of the high church party under Laud. It still however remained the intellectual centre of the puritan party; and of the numerous university graduates who emigrated to America between 1620 and 1647 over three-fourths came from Cambridge.

The moderate puritanism which had been predominant among the junior members of the university for a century and a half, and the moderate anglicanism which the majority of the senior members had professed for the same time, alike almost disappeared[1] with the excesses and violence in which the Independents indulged in the middle of the seventeenth century.

With the accession of Charles II. the same difference of opinion which had marked the Oxford and Cambridge of the reigns of Henry and Elizabeth again shewed themselves. Oxford adopted the anglicanism of Laud, and the politics of the extreme tories. Cambridge, on the other hand, gave rise to the school now known as that of the Cambridge Platonists, and was the centre of the whig party. I gather from Mullinger's work that the leading members of the Platonic school were Whichcote, Cudworth, Henry More, Culverwell, Rust, Glanvil, and Norris: they form the successors to the puritan divines of an earlier generation. The Platonists were succeeded in natural sequence by the school of Sherlock, Law, and Paley. They in their turn gave place on the one side to the evangelical school of Berridge, Milner, and Simeon; and on the other side, but somewhat later, to the school of Maurice, Trench, and Hallam.

External politics did not play so large a part in the internal history of the university as was the case at Oxford. Cambridge was the centre of the constitutional royalists at the

[1] See for example Pepys's diary for February 1660.

beginning of the sixteenth century, and of the whig party at the close of that century. The revolution of 1688 was the triumph of the latter. Towards the latter half of the eighteenth century the politics of the majority of the residents became tory rather than whig, but the toryism was of a moderate and progressive type.

In fact, both in religion and politics, the dominant tone of the university was what its friends would call moderation, tolerance, and a respect for the rights of others, and what its opponents would, I suppose, describe as lukewarmness, and a failure to carry principles to all their logical consequences.

The studies prevalent at the two universities mark the same difference of attitude[1]. At Oxford dogmatic theology, classical philosophy, and political history occupied most attention. At Cambridge the negative and critical philosophy and logic of Ramus was followed by the philosophy of Bacon (and possibly of Descartes), which in turn was displaced by that of Locke. The modern school of classical literature was worthily represented by Bentley, Porson, and others.

But it was the mathematical school which displayed the most marked originality and power. The writings of Briggs, Horrox, Wallis, Barrow, Newton, Cotes, and Taylor had placed Cambridge in the first rank of European schools. Under the influence of the Newtonian philosophy mathematics gradually became the dominant study of the place, and for the latter half of this time the mathematicians controlled the studies of the university almost as absolutely as the logicians

[1] It is interesting to observe how persistently particular studies have been prevalent at each of the two universities. Leaving aside literature and theology (to which much attention was paid at both universities), we may say that interest at Oxford has always been specially centred in philosophy in its wider sense, and history (constitutional and political); while at Cambridge the study of mathematical, physical, and natural science, and the applications thereof, have generally attracted more attention. Of course it is easy to cite particular instances to the contrary, but I believe the assertion above made is substantially true, and has been so for the last four hundred years.

had controlled those of the mediæval university. There can be no doubt that this was a real misfortune, and that it led to a certain one-sidedness in education. At the same time it must be remembered that a knowledge of the elements of moral philosophy and theology, an acquaintance with the rules of formal logic, and the power of reading and writing scholastic Latin were required from all students.

The mathematicians, to do them justice, threw no obstacle in the way of the introduction of other branches of learning; and the predominance of mathematical studies was mainly due to the fact that they were the only ones in which any continuous and conspicuous intellectual activity was displayed.

The isolation of the Cambridge mathematical school and the falling-off in the quality of the work produced are the most striking points in its position at the end of the last century. The adoption of the continental notation, the development of analytical methods, and the removal of the barriers which separated Cambridge mathematicians from their contemporaries of other schools distinguish the opening years of this century. Those reforms may be taken as effected by 1825. The achievements of the mathematical school for the years subsequent to that will form a brilliant chapter in the intellectual history of the university, but those who created the new school are too near our own time to render it possible or desirable to analyse the general characteristics of their work.

It was not however only in mathematics that this new renaissance was visible. In all branches of learning there was an awakening, and the last few years in which the Elizabethan statutes were in force are distinguished by the opening out of fresh studies, no less than by the development of old ones. Thus the year 1858 is the close of a well-defined period in the history of the university, and the new constitution then given to the university marks the beginning of another era, which I prefer to treat as wholly outside the limits of this work.

INDEX[1].

[1] The Index has been prepared at the University Press. I have revised and added to it and hope there are no omissions of importance. W. W. R. B.

CAMBRIDGE: PRINTED BY C. J. CLAY, M.A. AND SONS, AT THE UNIVERSITY PRESS.

Printed in the United States
By Bookmasters